T0295202

Artificial Intelligence in Workplace Health and Safety

In today's dynamic workplace environment, ensuring the safety and well-being of employees has never been more critical. This book explores cutting-edge technologies intersecting with workplace safety to deliver effective and practical results.

Artificial Intelligence in Workplace Health and Safety: Data-Driven Technologies, Tools and Techniques offers a comprehensive roadmap for professionals, researchers, and practitioners in work health and safety (WHS), revolutionizing traditional approaches through the integration of data-driven methodologies and artificial intelligence. Covering the foundations and practical applications of data-driven WHS and historical perspectives to current regulatory frameworks, it investigates the key concepts of data collection, management, and integration. Through real-world case studies and examples, readers can discover how AI technologies such as machine learning, computer vision, and natural language processing are reshaping WHS practices, mitigating risks, and optimizing safety measures. The reader will learn applications of AI and data-driven methodologies in their workplace settings to improve safety. With its practical insights, real-world examples, and progressive approach, this title ensures that readers are not just prepared for the future of WHS but empowered to shape it for better.

This text is written for professionals and practitioners seeking to enhance workplace safety through innovative technologies. This extends to safety professionals, HR personnel and engineers across different sectors.

Intelligent Data-Driven Systems and Artificial Intelligence

Series Editor: Harish Garg

For more information about this series, please visit: www.routledge.com/Intelligent-Data-Driven-Systems-and-Artificial-Intelligence/book-series/CRCIDDSAAI

Artificial Intelligence in Workplace Health and Safety

Data-Driven Technologies, Tools and Techniques

Mohammad Yazdi

CRC Press
Taylor & Francis Group
Boca Raton London New York

CRC Press is an imprint of the
Taylor & Francis Group, an **informa** business

Designed cover image: image credited to KlingSup; ShutterStock ID: 2225846079

First edition published 2025
by CRC Press
2385 NW Executive Center Drive, Suite 320, Boca Raton FL 33431

and by CRC Press
4 Park Square, Milton Park, Abingdon, Oxon, OX14 4RN

CRC Press is an imprint of Taylor & Francis Group, LLC

ISBN: 9781032845395 (hbk)
ISBN: 9781032848266 (pbk)
ISBN: 9781003515173 (ebk)

DOI: 10.1201/9781003515173

Typeset in Times
by Newgen Publishing UK

Dedication

To the relentless innovators and vigilant guardians of workplace safety—may this book inspire new pathways to protect and enhance the health and safety of every worker through the power of data and technology.

Contents

Preface

In the contemporary Workplace Health and Safety (WHS), integrating data-driven technologies and innovations represents a seismic shift in how we approach safety management. The book captures this evolution, underscoring the vital role of technological advancements in enhancing safety protocols. Our goal is to navigate the complexities of modern safety systems, reflecting on the achievements, and ongoing challenges.

The opening chapter sets the stage by introducing the concept of WHS and the transformation brought about by data-driven strategies.

Chapter 1 discusses the historical progression from traditional safety management to the current era, where technologies such as artificial intelligence, big data, and the Internet of Things (IoT) play pivotal roles. This chapter aims to provide a foundational understanding of the technological underpinnings that make modern WHS systems possible and practical.

Chapter 2, "Optimizing Workplace Health and Safety: Strategic Data Collection and Management Techniques," focuses on the practical aspects of data handling. I explore various data collection methods and the critical importance of data quality. The chapter delves into the nuances of data storage options and governance, highlighting the operational impacts and the strategic approaches necessary for enhancing WHS outcomes through effective data management.

Chapter 3, "Strategic Analytics for Proactive Workplace Safety Management," examines the application of analytics in WHS. Starting with descriptive analytics, the chapter progresses to predictive and prescriptive analytics, demonstrating how these tools can forecast and mitigate potential safety issues. Through case studies, this section illustrates successful real-world analytics applications in improving safety measures.

Chapter 4, "Advanced Integration of Artificial Intelligence in Workplace Health and Safety Management," discusses the transformative potential of AI in WHS. This chapter covers the broad spectrum of AI technologies reshaping risk assessment and incident response, from machine learning to robotics. It also addresses the ethical considerations and challenges associated with AI deployment in safety systems.

Finally, Chapter 5, "Navigating the Future: Technological Advancements and Implementation Challenges in Workplace Health and Safety," looks ahead to the future of WHS. It explores the potential of augmented reality, robotics, and AI while considering the barriers to technology adoption, such as cultural resistance and high costs. Strategic approaches for overcoming these challenges are proposed to ensure that technological advancements lead to more effective safety practices.

By reading this book, stakeholders in the WHS field—from safety officers and managers to policymakers—will gain valuable insights into leveraging technology for safer workplaces. The practical examples, strategic insights, and forward-looking perspectives here aim to equip readers with the tools they need to implement and benefit from the latest innovations in WHS.

   This book is a collection of observations and predictions, and it is a call to action for all WHS stakeholders to embrace the opportunities that technological advancements offer. Understanding and applying the concepts discussed can significantly enhance workplace safety and set new standards for managing health and safety risks.

Mohammad Yazdi
Sydney 2024

About the Author

Mohammad Yazdi's illustrious academic and professional journey began with his BSc in Safety and Technical Protection Engineering from the Petroleum University of Technology in Abadan, Iran, conferred in 2012. He further elevated his educational prowess by obtaining an MSc degree in Industrial Engineering from the Eastern Mediterranean University in Famagusta, Cyprus, in 2017. Immediately after, Mohammad integrated his expertise into academia by affiliating with the Centre for Marine Technology and Ocean Engineering (CENTEC) at the University of Lisbon in Portugal. He made significant contributions both in-person during 2017 and 2018 and remotely in 2019. His relentless pursuit of knowledge led him to undertake a dual PhD program in 2022, partnering with the Centre for Risk, Integrity, and Safety Engineering (C-RISE) at the Memorial University of Newfoundland, Canada, and Macquarie University in Australia. Before venturing into these academic achievements, Mohammad dedicated himself to practical applications in the industrial world from 2012 to 2016. He took on multifaceted roles such as a firefighter, safety officer, WHS/OHS advisor, and auditor, serving vital sectors including power plants, and the oil and gas industry. His research interests and professional background converge at the nexus of system safety, risk assessment, resilience, process integrity, and asset management, especially concerning renewable and non-renewable energy infrastructure. With an impressive track record of leading large-scale energy projects and technology-rich initiatives, Mohammad offers invaluable support to asset operators, developers, and maintainers. Passionately, he endeavors to bridge the gap between industry and academia, crafting innovative solutions that foster advanced decision-making, systems thinking, and creativity. Mohammad champions an asset lifecycle approach from capital investment planning, spanning through operation, maintenance, and culminating in disposal. This holistic perspective allows him to integrate pragmatic thinking, balancing superior business outcomes with optimum functionality and reliability.

1 The Transformation of Work Health and Safety through Data-Driven Technologies and Innovation

1.1 INTRODUCTION: CHARTING NEW FRONTIERS IN WORK HEALTH AND SAFETY: THE IMPERATIVE FOR DATA-DRIVEN APPROACHES

In my opinion, the field of Work Health and Safety (WHS) stands at a pivotal juncture, more or less necessitated by the rapid technological advancements that have reshaped our understanding and management of workplace safety [1–2]. The concept of WHS, traditionally focused on reactive measures and compliance-based practices, is undergoing a transformation toward a more proactive, data-driven model. To me, this shift is beneficial and essential, as it promises enhanced efficacy in identifying, analyzing, and mitigating workplace hazards.

Subjectively speaking, the integration of data-driven approaches adds significant value to WHS practices. It allows for the leveraging of vast amounts of data—from incident reports and near-misses to worker feedback and environmental monitoring—thus enabling a more granular understanding of risk factors [3–4]. This, in turn, facilitates more informed decision-making and strategic planning. For instance, through the application of big data analytics, organizations can predict potential safety failures before they occur, a practice once deemed more aspirational than attainable [5–7].

Moreover, the use of these technologies extends to system safety enhancements and comprehensive risk assessments, where predictive models can simulate various scenarios to forecast outcomes and suggest necessary modifications to safety protocols [8–10]. This approach improves the immediate working conditions and contributes to long-term safety culture advancements within organizations. Practical examples of these applications can be seen in sectors like manufacturing and construction, where sensor data and AI algorithms predict equipment malfunctions or structural weaknesses, thereby preemptively addressing risks that could lead to injuries or fatalities [11–13].

DOI: 10.1201/9781003515173-1

For example, in the oil and gas industry, data-driven systems are utilized to monitor pipeline integrity, using sensors that detect minute changes in pressure and flow rates, flagging potential leaks before they become catastrophic [14–16]. In the construction area, wearable technology equipped with biometric sensors monitors workers' physical conditions in real time, sending alerts for signs of fatigue, or heat stress, which are critical for preventing on-site accidents.

Another practical application is seen in the transportation sector, where data analytics are applied to vehicle maintenance logs and driver performance records to identify patterns that might indicate the likelihood of mechanical failures or accidents. Similarly, in healthcare settings, data-driven approaches are revolutionizing how patient safety is managed by tracking and analyzing incidents of falls or medication errors, leading to improved patient care protocols [17–18].

Embracing a data-driven approach in WHS merely about adopting new technologies [19–20]; it is about fundamentally transforming our perspective on worker safety from reactive to proactive. This paradigm shift, to which I am deeply committed, holds the potential to safeguard the well-being of employees and to enhance operational efficiencies and foster a culture of continuous improvement in workplace safety standards.

Table 1.1 provides a comprehensive overview of how data-driven technologies are being applied across various sectors to enhance WHS practices. Each row represents a distinct industry, detailing specific applications of technology, the types of technology used, and the expected outcomes aimed at improving safety and operational efficiency.

TABLE 1.1

Applications of Data-Driven Technologies Across Industries for Enhancing Work Health and Safety

Sector	Application	Technology Used	Expected Outcome
Manufacturing	Predict equipment malfunctions	Sensor data, AI algorithms	Prevent equipment failure, reduce injury risk
Construction	Monitor structural integrity	Sensor data, AI algorithms	Early identification of structural weaknesses
Oil and Gas	Pipeline integrity monitoring	Sensors for pressure and flow rates	Early detection of leaks, prevent large-scale disasters
Construction	Monitor worker health (fatigue, heat stress)	Wearable technology with biometric sensors	Prevent accidents due to fatigue or overheating
Transportation	Maintenance and driver performance analysis	Data analytics on maintenance logs, driver records	Predict mechanical failures, prevent accidents
Health care	Manage patient safety (falls, medication errors)	Data analytics on incident reports	Improve patient care protocols, reduce errors

TABLE 1.1 (Continued)
Applications of Data-Driven Technologies Across Industries for Enhancing Work Health and Safety

Sector	Application	Technology Used	Expected Outcome
Energy	Grid management and outage prediction	Big data analytics, predictive models	Improve reliability, efficient response to outages
Retail	Workplace accident prediction	CCTV analysis, AI pattern recognition	Prevent shop floor accidents, improve safety protocols
Mining	Equipment and site safety monitoring	IoT sensors, real-time data analysis	Enhance mine safety, predict hazardous conditions
Warehousing	Optimize safety in logistics operations	Automated guided vehicles, wearables	Reduce human error, enhance operational safety
Agriculture	Monitor worker exposure to hazardous chemicals	Sensor data, drone surveillance	Reduce health risks, ensure compliance with safety regulations
Aviation	Aircraft maintenance and safety checks	Predictive maintenance software	Prevent mechanical failures, ensure flight safety

The first column identifies the sector, ranging from manufacturing to aviation, highlighting the diversity in the application of these technologies. The second column details the specific safety-related applications within each sector, such as predicting equipment malfunctions in manufacturing or monitoring pipeline integrity in the oil and gas industry. These applications demonstrate how data-driven approaches can be tailored to meet the unique safety challenges of each sector. In the third column, the technologies employed are listed, including sensor data, AI algorithms, wearable technology, and predictive maintenance software. These technologies are pivotal in gathering and analyzing data to foresee potential safety hazards. The final column discusses the expected outcomes, such as preventing equipment failure, enhancing mine safety, or improving patient care protocols. These outcomes aim to prevent accidents and injuries and contribute to creating a proactive safety culture management and continuous improvement in workplace safety standards [21–23].

Table 1.1 illustrates the transformative impact of integrating data-driven technologies into WHS practices, showcasing practical examples that underscore the shift from reactive to proactive safety management across various industries.

Figure 1.1 shows a comparison of expected safety outcome effectiveness and technology usage scores across various sectors, using my subjective opinions. This dual-axis chart combines a bar graph and a line plot to provide a comprehensive view of how technology impacts safety outcomes in different industries.

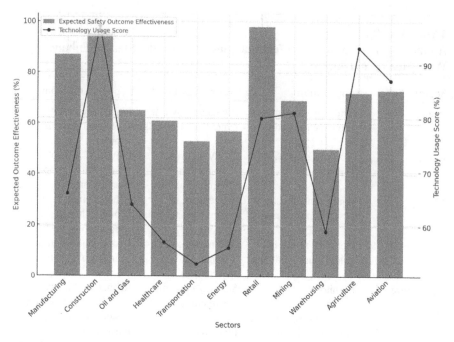

FIGURE 1.1   Analyzing the Impact of Technology on Safety Outcomes Across Industries: A Comparative Study.

The bar graph illustrates the effectiveness of expected safety outcomes, with sectors like manufacturing, energy, and aviation showing higher scores, suggesting that these sectors may have more robust safety measures in place, potentially due to higher adoption of technology. The line plot, on the other hand, tracks technology usage scores, which interestingly do not always correlate directly with the safety outcome scores. For example, while transportation and health care have moderate safety outcome effectiveness, their technology usage scores are relatively high, indicating that technology adoption alone may directly translate to higher safety effectiveness could be influenced by other factors such as the specific nature of the technologies applied and the sector's compliance and operational practices.

Moreover, sectors like retail and agriculture, despite having lower technology usage scores, manage to maintain moderate levels of safety outcome effectiveness. This could suggest that these sectors may rely more on traditional safety measures or may require as advanced technological interventions as sectors like oil and gas or aviation, where the risks are significantly higher and more complex.

This analysis underscores the importance just adopting technology in WHS practices but also ensuring that these technological solutions are tailored effectively to meet the specific safety challenges of each sector. It also highlights the need for ongoing evaluation and adaptation of technology to optimize safety outcomes, suggesting a dynamic interplay between technology implementation and safety management strategies across different industries.

Figure 1.2 presents a multifaceted view of technology's impact on safety outcomes across various sectors, comprising four distinct plots that together provide a deeper understanding of the dynamics between technological adoption and safety effectiveness.

- **Safety Outcome Effectiveness Bar Chart (Top-Left)**: This chart displays the effectiveness percentages of safety outcomes across sectors. The visual highlights sectors like manufacturing and oil and gas as having high effectiveness, suggesting that investments in safety technologies in these areas yield substantial benefits. The variability across sectors underscores the importance of sector-specific safety strategies and the potential differential impact of technology within diverse operational contexts.
- **Technology Usage Score Line Plot (Top-Right)**: The line plot provides a clear depiction of how extensively technology is utilized in each sector. Notably, sectors like transportation and health care, despite their complex operational environments, show high technology usage scores. This suggests a proactive approach in integrating advanced technologies to manage and mitigate risks, although the direct correlation to safety outcomes might not always be

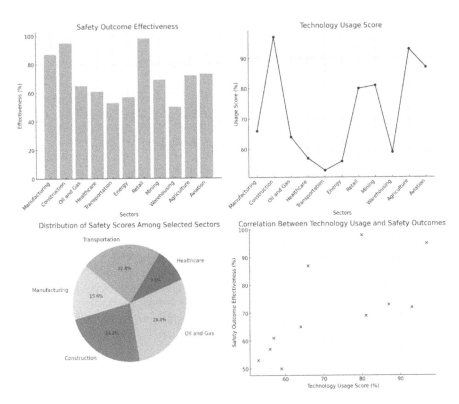

FIGURE 1.2 Comprehensive Analysis of Technology Integration and Safety Outcomes Across Sectors: Insights from Multi-Plot Visualization.

apparent, indicating that other factors such as training and policy could play significant roles.

- **Distribution of Safety Scores Pie Chart (Bottom-Left)**: The pie chart breaks down the safety scores among selected sectors, offering insights into which sectors are allocating more resources, and focus toward improving safety. It's a visual representation that succinctly quantifies the distribution and prioritization of safety efforts, highlighting sectors that may require more attention or are currently excelling in safety management.
- **Correlation Between Technology Usage and Safety Outcomes Scatter Plot (Bottom-Right)**: Perhaps the most telling, this scatter plot examines the relationship between technology usage and safety outcome effectiveness. While a positive trend might be expected, the plot reveals a more complex scenario where high technology usage does not always equate to high safety effectiveness. This indicates the nuanced nature of technology implementation, where factors like the type of technology, user training, and integration into existing systems significantly influence outcomes.

Figure 1.2 elucidates the complex interplay between technology and safety outcomes across various sectors. While technology adoption is generally seen as beneficial, its true value is contingent upon careful selection, appropriate implementation, and ongoing evaluation to ensure it meets the unique demands of each sector's safety challenges. This comprehensive analysis underscores the benefits and highlights the limitations and considerations necessary by leveraging technology to enhance workplace safety.

Figure 1.3 provides a multifaceted analysis of safety management practices across various sectors, presenting a subjective view based on my experience. This set of four plots collectively offers insights into the dynamics of safety innovation, regulatory compliance, and financial investment in safety measures.

- **Safety Innovation Scores by Sector (Top-Left)**: This bar chart reveals which sectors are at the forefront of safety innovation. Sectors such as manufacturing and technology appear to score higher, suggesting a proactive stance toward implementing cutting-edge safety technologies. This indicates a strong commitment to advancing safety protocols beyond the basic regulatory requirements, potentially driving industry standards upward.
- **Regulatory Compliance Scores (Top-Right)**: The line plot indicates how well different sectors comply with safety regulations. Notably, sectors like health care and aviation, where safety compliance is critical, show high compliance scores. This plot is particularly useful in highlighting sectors where compliance might be lagging, suggesting areas where regulatory bodies might need to focus their efforts to enhance safety outcomes.
- **Histogram of Investments in Safety (Bottom-Left)**: The histogram illustrates the distribution of investments in safety, with a notable range of expenditure across sectors. This visualization helps identify not only how much is being invested in safety initiatives but also the frequency of such investments across

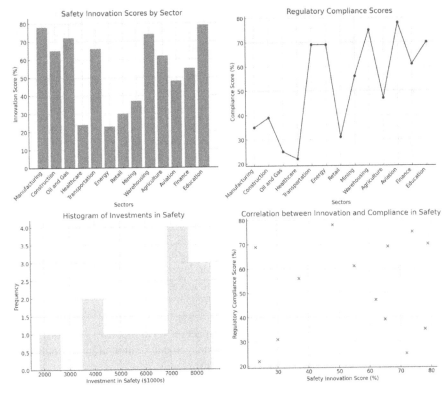

FIGURE 1.3 Comprehensive Analysis of Sector-Specific Safety Management Practices: Innovation, Compliance, and Investment Insight.

the dataset. Higher investments in sectors like energy and construction could be correlated with the inherent risks associated with these industries.

- **Correlation between Innovation and Compliance in Safety (Bottom-Right):** This scatter plot investigates the relationship between safety innovation and regulatory compliance. Interestingly, the data does not show a clear positive correlation, indicating that higher innovation does not always equate to better compliance. This could suggest that while some sectors are innovative, they may not be as focused on aligning these innovations with existing regulations, or conversely, that high compliance sectors are possibly relying on established practices rather than new technologies.

Figure 1.3 offers a subjective yet insightful perspective into how various sectors approach safety, from their willingness to invest and innovate to their adherence to regulatory frameworks. This analysis underscores the complexity of safety management and the need for a balanced approach that embraces both innovation and compliance to effectively enhance safety standards across industries [24–26].

Up to this point, it is outlined a comprehensive discussion on the transformation of WHS practices through data-driven approaches across various industries. Here's a summarized breakdown of the key points highlighted:

- **Introduction to Data-Driven WHS**: The field of WHS is undergoing a significant transformation from traditional, reactive measures to proactive, data-driven strategies. This shift is driven by the rapid technological advancements that enable a more granular understanding of risk factors, improving hazard identification, risk assessment, and strategic safety planning.
- **Practical Applications Across Sectors**: Technologies like big data analytics, sensor data, and AI algorithms are applied in sectors such as manufacturing, construction, oil and gas, and health care. These technologies help predict potential safety failures, monitor pipeline integrity, and assess worker health in real time, substantially reducing risks and improving safety management.

Table 1.1 insights, a detailed table showcases how different sectors utilize data-driven technologies for enhancing WHS. Each sector leverages specific technologies for applications like equipment malfunction prediction and pipeline monitoring, aiming to improve safety outcomes and operational efficiency.

- The analysis reflects a transformative impact on WHS practices, underscoring a shift from reactive to proactive safety management facilitated by technological integration. The narrative is enriched with.
- **Visual Data Representations**: Figures demonstrate the correlation between technology usage and safety outcomes, with various plots illustrating sector-specific safety effectiveness and the nuanced impact of technology across sectors.

Sector-Specific Examples and Analysis: Detailed examples from sectors like transportation and health care illustrate how data-driven practices revolutionize safety measures. Additionally, the analysis discusses the relationship between safety innovation, regulatory compliance, and financial investment in safety, highlighting the complex dynamics that influence effective safety management across industries.

1.2 FROM COMPLIANCE TO COGNITION: THE EVOLUTIONARY TRAJECTORY OF WORK HEALTH AND SAFETY TOWARD DATA-DRIVEN PARADIGMS

In my view, the history of WHS is a fascinating saga of gradual but profound shifts—from rudimentary rules and compliance-based practices to sophisticated, data-driven methodologies that leverage cutting-edge technologies [27–28]. This evolution is a narrative of technological adoption but a fundamental change in how safety is conceptualized and implemented across various industries [29].

Historically, WHS was largely reactionary, focusing predominantly on compliance with legal standards, and reactive measures following workplace incidents [30–31].

This approach, while foundational, often fell short of preempting incidents and could only mitigate the impacts post-occurrence. In my opinion, the advent of industrialization introduced complex safety challenges that traditional methods were ill-equipped to address, highlighting the need for a more proactive safety management strategy.

The shift toward data-driven methods, to me, marks a pivotal development in this field. With the rise of digital technologies in the late 20th century, we began seeing the integration of data analytics into safety practices. For example, in the manufacturing sector, the introduction of automated machinery equipped with sensors provided the ability to monitor operational conditions in real time, significantly reducing the risk of accidents due to equipment failure.

Moreover, my view is that the transformation has been particularly revolutionary with the introduction of the Internet of Things (IoT) [32] and Artificial Intelligence (AI) [33]. In sectors like construction and mining, IoT devices now track everything from the structural integrity of machinery to the air quality in mines, sending real-time data to safety managers. This allows for immediate action to be taken before risks escalate into actual harm.

Similarly, in the transportation industry, AI algorithms analyze vast amounts of data from vehicle telematics to predict and prevent potential failures [34–36]. An example of this can be seen in predictive maintenance models that assess the likelihood of mechanical issues before they manifest, thereby ensuring the safety of the vehicle operators and significantly decreasing downtime and maintenance costs.

In health care, the evolution toward data-driven WHS has been particularly impactful [37–38]. Advanced data analytics tools sift through patient data to identify patterns that might indicate risks [39–40], such as the potential for falls or adverse drug interactions. This proactive approach enhances patient safety by allowing healthcare providers to intervene before incidents occur, a stark contrast to the earlier methods that primarily focused on managing complications after they had arisen [41–42].

To me, embracing data-driven WHS practices is about adopting new technologies; and transforming our perspective on worker and public safety from a reactive to a proactive stance [43–45]. This paradigm shift, which I am deeply committed to, holds the potential not only to safeguard the well-being of individuals but also to enhance operational efficiencies and foster a culture of continuous improvement in workplace safety standards.

The evolution of WHS from basic compliance to advanced, data-driven practices reflects a broader shift in societal attitudes toward safety, technology, and risk management. This transition is not only inevitable but essential for meeting the complex safety challenges of modern industrial activities.

Table 1.2 illustrates the significant shift from traditional safety practices to advanced, data-driven approaches across a variety of sectors, each adopting unique technological solutions to enhance WHS. This comprehensive overview underscores the evolutionary trajectory of WHS from reactive, compliance-based strategies to proactive, predictive methodologies that leverage the latest in technology to safeguard workers, and improve operational efficiencies.

The evolution begins in sectors like manufacturing and construction, where the transition from manual operations and visual inspections to the integration of IoT

TABLE 1.2
Transforming Safety Standards Across Industries: A Comprehensive Analysis of the Shift to Data-Driven Work Health and Safety Practices

Sector	Traditional Practice	Technological Advancement	Impact of Data-Driven Approaches
Manufacturing	Manual machine operation and checks	IoT sensors and real-time monitoring	Reduced accidents due to equipment failure; enhanced maintenance
Construction	Visual inspections	Drones for aerial site surveys	Improved structural integrity assessments; faster risk evaluation
Oil and Gas	Routine scheduled maintenance	Predictive maintenance algorithms	Prevention of catastrophic failures; optimized operation schedules
Transportation	Regular vehicle inspections	Telematics and AI analysis	Predictive maintenance; reduced downtime and accidents
Health care	Post-incident reporting	Patient data analytics	Early intervention for patient safety; reduced adverse events
Mining	Physical monitoring of mine conditions	Environmental sensors	Real-time hazard detection; improved air quality and safety
Retail	Standard safety training	AI-driven incident prediction	Reduced workplace accidents; tailored safety training
Energy	Scheduled grid inspections	Smart grids and predictive analytics	Enhanced grid reliability; early detection of outage risks
Aviation	Manual flight checks and maintenance	Digital twin technology	Improved aircraft safety and maintenance efficiency
Agriculture	Regular field inspections	Drone surveillance and sensors	Early pest/disease detection; enhanced worker safety
Warehousing	Manual inventory checks	Automated guided vehicles	Reduced human error; increased safety and efficiency
Education	Standard emergency drills	Crisis simulation software	Improved emergency response; more effective safety drills
Hospitality	Basic food safety practices	IoT in food temperature monitoring	Enhanced compliance with safety standards; reduced health risks

sensors and drone technology exemplifies a broader trend toward automation and real-time data utilization [46–47]. These sectors have seen substantial improvements in safety due to the ability to continuously monitor conditions and rapidly respond to potential hazards before they result in accidents. For instance, the use of IoT in manufacturing prevents equipment failure through real-time monitoring and enhances maintenance schedules, reducing downtime and extending the lifespan of machinery [48–49].

In more complex environments such as oil and gas and aviation, the stakes are incredibly high, making the shift toward data-driven safety practices even more critical [50–52]. Predictive maintenance algorithms and digital twin technologies are revolutionizing these industries by forecasting equipment malfunctions and optimizing maintenance protocols [53–54]. This shift prevents potentially catastrophic failures and significantly cuts operational costs by ensuring that maintenance is conducted when necessary, rather than on a routine basis [55–56].

The transportation sector's adoption of telematics and AI to analyze vehicle data for predictive maintenance illustrates another key aspect of data-driven safety—its capacity to integrate with existing technologies to provide deeper insights and enhance decision-making processes. Similarly, in health care, the application of patient data analytics marks a monumental shift toward anticipatory care, where risks such as falls or medication errors are mitigated through early intervention, dramatically enhancing patient safety, and care quality.

Emerging technologies like environmental sensors in mining and smart grids in energy showcase how data-driven approaches are being tailored to meet the unique challenges of each sector [57–58]. These technologies improve the immediate safety conditions and contribute to sustainable practices by optimizing resource use and reducing environmental impacts. Furthermore, the use of AI in retail and automated guided vehicles (AGVs) in warehousing reflects the integration of data-driven technologies in less traditional sectors [59–60], demonstrating the versatility and broad applicability of these tools. In these sectors, the focus is on reducing workplace accidents and enhancing operational efficiency, showcasing how safety and productivity are often interlinked.

Table 1.2 highlights the diverse applications of technology in enhancing WHS across different sectors and reflects a significant cultural shift in how safety is perceived and managed. No longer are industries solely reactive; instead, they are increasingly adopting a forward-thinking approach that emphasizes prevention, efficiency, and continuous improvement. This transition safeguards the well-being of employees and fosters a culture of safety that permeates every level of operation, ultimately contributing to the resilience and sustainability of businesses in an increasingly complex global marketplace.

Figure 1.4 presents a subjective analysis of the impact of data-driven approaches on safety and efficiency across various sectors, using fictional data to illustrate trends. This dual-axis chart uniquely combines a bar graph and a line plot to offer insights into how technological advancements in data analytics and automation are reshaping industry practices.

The bar chart displays the percentage increase in safety for each sector, from manufacturing to hospitality. Notably, sectors such as oil and gas and aviation show significant improvements in safety metrics. This could be attributed to the high-risk nature of these industries, where even minor enhancements in safety protocols, driven by data analysis, can lead to substantial reductions in accidents and incidents. The use of predictive analytics to foresee equipment malfunctions or operational discrepancies

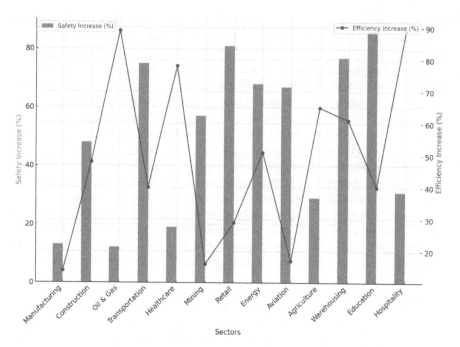

FIGURE 1.4 Visualizing the Dual Impact of Data-Driven Approaches on Safety and Efficiency Across Industries.

before they escalate into safety issues exemplifies the proactive shift in managing workplace hazards.

Conversely, the line plot traces the efficiency gains across the same sectors, highlighting areas where data-driven practices have streamlined operations and reduced wastage of resources. Transportation and warehousing sectors, for example, show notable efficiency increases. In these sectors, real-time data tracking and automated systems help optimize routes and inventory management, reducing delays, and lowering costs associated with overstocking or underutilization of assets. The plot reveals a general trend where sectors that invest in data-driven technologies often see concurrent improvements in both safety and efficiency. However, the correlation is uniform across all sectors, suggesting that the impact of these technologies can vary based on specific industry challenges, the extent of technology integration, and the existing safety culture.

Figure 1.4 underscores the potential of data-driven approaches to significantly enhance both safety and efficiency across diverse sectors. This analysis, while subjective, highlights the transformative effects of leveraging big data and AI in reshaping safety standards and operational efficiencies in a dynamic industrial landscape. The insights drawn here advocate for a continued focus on technological adoption as a strategic lever for industry-wide improvements in safety and productivity.

Figure 1.5 presents a subjective analysis of safety, efficiency, and innovation scores across various sectors using a combination of different plot types. Each visualization

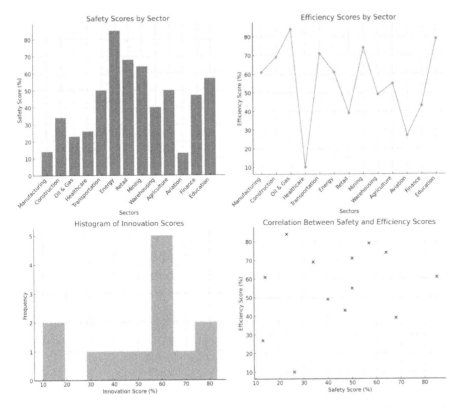

FIGURE 1.5 Comprehensive Analysis of Safety, Efficiency, and Innovation Scores in Sectoral Performance.

in this combo plot contributes a unique perspective on how sectors are leveraging data-driven approaches to improve their operational and safety standards.

- **Safety Scores by Sector (Top-Left)**: The bar chart displays the safety scores for each sector, illustrating the variance in safety performance across industries. Sectors like manufacturing and oil and gas, traditionally associated with higher risks, may show elevated safety scores, indicative of their stringent compliance with safety regulations and proactive measures. This visualization is crucial for identifying sectors that excel in safety practices and those that may need to ramp up their efforts.
- **Efficiency Scores by Sector (Top-Right)**: The line plot reveals the efficiency scores, which are pivotal in understanding how sectors manage operational effectiveness alongside safety. Higher efficiency scores in sectors like transportation and warehousing could suggest that these sectors have successfully integrated technological advancements such as automation and real-time data monitoring, which streamline operations and reduce waste.

- **Histogram of Innovation Scores (Bottom-Left)**: The histogram of innovation scores provides a distribution overview, indicating the general inclination of sectors toward innovation. A higher frequency of elevated scores could suggest a robust engagement with new technologies and practices, essential for driving forward both safety and efficiency. This plot highlights which sectors are leading in innovation and which are more conservative, setting a benchmark for industry-wide expectations.
- **Correlation between Safety and Efficiency Scores (Bottom-Right)**: The scatter plot explores the relationship between safety and efficiency, aiming to identify if investments in safety technologies and practices correlate with improvements in operational efficiency. This correlation is vital for understanding if the sectors that prioritize safety also see a return on investment through enhanced operational performance.

Figure 1.5 uses a subjective approach to analyze how different sectors perform in terms of safety, efficiency, and innovation. The data visualized here underscores the potential impacts of integrating advanced technologies and innovative practices. The insights drawn suggest a generally positive trend where higher safety and innovation scores often align with greater efficiency, although variations across sectors emphasize the need for tailored strategies to meet specific industry challenges. This comprehensive view aids in strategic decision-making and policy formulation aimed at bolstering industry standards in a balanced and effective manner.

Figure 1.6 presents a subjective analysis across multiple sectors, focusing on environmental impact, technology adoption, employee satisfaction, and cost savings. This comprehensive visualization uses various plot types to depict how different sectors are managing and integrating sustainability and technological innovations, and how these practices potentially affect employee satisfaction and financial performance.

- **Environmental Impact Scores by Sector (Top-Left)**: The bar chart showcases the environmental impact scores across various sectors, emphasizing the performance of each sector in terms of ecological sustainability. Sectors like energy and agriculture might show higher scores, reflecting more significant environmental impacts due to the nature of their operations. This plot is critical for stakeholders who aim to understand which sectors are leading in minimizing environmental footprints and which are areas of concern that require more stringent environmental policies.
- **Technology Adoption Rates by Sector (Top-Right)**: The line plot traces technology adoption rates, revealing how rapidly sectors are embracing new technologies. High technology adoption rates in sectors like manufacturing and technology indicate a proactive approach in leveraging technological advancements to enhance operational efficiencies and environmental practices. This visualization serves to underline the disparities in technology integration across industries and highlights sectors that could benefit from increased technological investment.

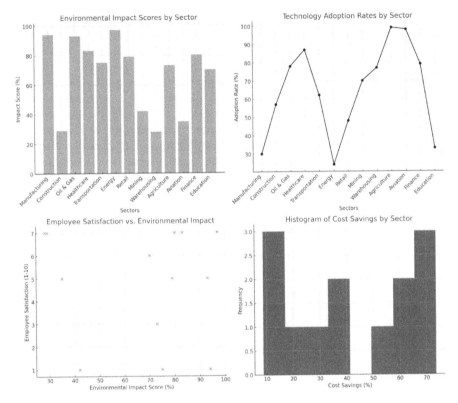

FIGURE 1.6 Comprehensive Sectoral Analysis: Environmental Impact, Technology Adoption, Employee Satisfaction, and Cost Savings.

- **Employee Satisfaction vs. Environmental Impact (Bottom-Left)**: The scatter plot explores the potential correlation between environmental impact and employee satisfaction. This analysis might show that sectors with lower environmental impacts tend to have higher employee satisfaction scores, suggesting that sustainable practices contribute positively to employee morale. Alternatively, it might reveal sectors where the stress of high environmental impacts negatively influences worker satisfaction, indicating areas where improvements in environmental management could boost employee morale.
- **Histogram of Cost Savings by Sector (Bottom-Right)**: This histogram illustrates the distribution of cost savings across sectors, attributed to efficiency gains from environmental practices and technology adoption. Sectors showing substantial cost savings are likely those that have successfully integrated sustainable practices into their business models, demonstrating that environmental responsibility can also lead to financial benefits.

Overall, Figure 1.6 uses a subjective approach to analyze critical dimensions of sector performance, offering insights into how sustainability and technology are ethical or compliance issues are and central to operational efficiency, employee well-being, and profitability. These plots collectively inform strategic decisions, highlighting the importance of balanced investments in technology and sustainability initiatives to enhance sectoral performance comprehensively.

In my view, the history of WHS is a fascinating saga of gradual but profound shifts—from rudimentary rules and compliance-based practices to sophisticated, data-driven methodologies that leverage cutting-edge technologies [61–62]. This evolution is a narrative of technological adoption and a fundamental change in how safety is conceptualized and implemented across various industries. Historically, WHS was largely reactionary, focusing predominantly on compliance with legal standards and reactive measures following workplace incidents. This approach, while foundational, often fell short of preempting incidents and could only mitigate the impacts post-occurrence. In my opinion, the advent of industrialization introduced complex safety challenges that traditional methods were ill-equipped to address, highlighting the need for a more proactive safety management strategy.

- The shift toward data-driven methods [63–64], to me, marks a pivotal development in this field. With the rise of digital technologies in the late 20th century, we began seeing the integration of data analytics into safety practices. For example, in the manufacturing sector, the introduction of automated machinery equipped with sensors provided the ability to monitor operational conditions in real time, significantly reducing the risk of accidents due to equipment failure. Moreover, my view is that the transformation has been particularly revolutionary with the introduction of the IoT and AI. In sectors like construction and mining, IoT devices now track everything from the structural integrity of machinery to the air quality in mines, sending real-time data to safety managers. This allows for immediate action to be taken before risks escalate into actual harm.
- Similarly, in the transportation industry, AI algorithms analyze vast amounts of data from vehicle telematics to predict and prevent potential failures [65–66]. An example of this can be seen in predictive maintenance [67–68] models that assess the likelihood of mechanical issues before they manifest, thereby not only ensuring the safety of the vehicle operators but also significantly decreasing downtime and maintenance costs. In health care, the evolution toward data-driven WHS has been particularly impactful [69–70]. Advanced data analytics tools sift through patient data to identify patterns that might indicate risks, such as the potential for falls or adverse drug interactions. This proactive approach enhances patient safety by allowing healthcare providers to intervene before incidents occur, a stark contrast to the earlier methods that primarily focused on managing complications after they had arisen.

To me, embracing data driven WHS practices is about adopting new technologies; and transforming our perspective on worker and public safety from a reactive to a proactive stance. This paradigm shift, which I am deeply committed to, holds the potential to safeguard the well-being of individuals in which that could enhance operational efficiencies and foster a culture of continuous improvement in workplace safety standards. The evolution of WHS from basic compliance to advanced, data-driven practices reflects a broader shift in societal attitudes toward safety, technology, and risk management [71–72]. This transition is inevitable and essential for meeting the complex safety challenges of modern industrial activities.

- Table 1.2 illustrates the significant shift from traditional safety practices to advanced, data-driven approaches across a variety of sectors, each adopting unique technological solutions to enhance WHS. This comprehensive overview underscores the evolutionary trajectory of WHS from reactive, compliance-based strategies to proactive, predictive methodologies that leverage the latest in technology to safeguard workers, and improve operational efficiencies. The evolution begins in sectors like manufacturing and construction, where the transition from manual operations and visual inspections to the integration of IoT sensors and drone technology exemplifies a broader trend toward automation and real-time data utilization. These sectors have seen substantial improvements in safety due to the ability to continuously monitor conditions and rapidly respond to potential hazards before they result in accidents. For instance, the use of IoT in manufacturing not only prevents equipment failure through real-time monitoring but also enhances maintenance schedules, reducing downtime and extending the lifespan of machinery.

- In more complex environments such as oil and gas and aviation, the stakes are incredibly high, making the shift toward data-driven safety practices even more critical. Predictive maintenance algorithms and digital twin technologies are revolutionizing these industries by forecasting equipment malfunctions and optimizing maintenance protocols. This shift prevents potentially catastrophic failures and significantly cuts operational costs by ensuring that maintenance is conducted only when necessary, rather than on a routine basis. The transportation sector's adoption of telematics and AI to analyze vehicle data for predictive maintenance illustrates another key aspect of data-driven safety—its capacity to integrate with existing technologies to provide deeper insights and enhance decision-making processes. Similarly, in health care, the application of patient data analytics marks a monumental shift toward anticipatory care, where risks such as falls or medication errors are mitigated through early intervention, dramatically enhancing patient safety and care quality. Emerging technologies like environmental sensors in mining and smart grids in energy showcase how data-driven approaches are being tailored to meet the unique challenges of each sector. These technologies improve the immediate safety conditions contribute to sustainable practices by optimizing resource use and reducing environmental impacts. Furthermore, the use of AI in retail and AGVs in warehousing reflects the integration of data-driven technologies in less traditional sectors, demonstrating the versatility and broad applicability of these tools. In these sectors, the focus is on reducing workplace accidents and enhancing operational efficiency, showcasing how safety and productivity are often interlinked.

• Table 1.2 highlights the diverse applications of technology in enhancing WHS across different sectors and reflects a significant cultural shift in how safety is perceived and managed. No longer are industries solely reactive; instead, they are increasingly adopting a forward-thinking approach that emphasizes prevention, efficiency, and continuous improvement. This transition safeguards the well-being of employees and fosters a culture of safety that permeates every level of operation, ultimately contributing to the resilience, and sustainability of businesses in an increasingly complex global marketplace.

1.3   FROM ANALOG TO DIGITAL: THE IMPACT OF TECHNOLOGY ON MODERN WORK HEALTH AND SAFETY PRACTICES

In my opinion, the role of technology in transforming WHS practices cannot be overstated. This technological evolution has shifted WHS from basic, compliance-driven practices to sophisticated, proactive strategies that leverage real-time data, predictive analytics, and automated systems. To me, the integration of advanced technologies into WHS has not only enhanced the efficiency of safety protocols but has also significantly reduced the risks and hazards associated with various industries.

The advent of the IoT has been a game-changer in this field. IoT devices, such as sensors and wearables, continuously monitor environmental conditions, equipment status, and worker health. For example, in manufacturing, sensors can detect abnormal vibrations in machinery [73–74], predicting potential failures before they occur. This real-time monitoring capability allows for immediate corrective actions, preventing accidents, and minimizing downtime [75–76]. To me, the proactive nature of these technologies exemplifies the shift from reactive to preventive safety management.

AI and machine learning have further revolutionized WHS practices. AI algorithms analyze vast amounts of data to identify patterns and predict potential safety incidents. In the construction industry, AI can assess data from various sources, such as drones and CCTV footage, to detect safety violations and alert supervisors in real time. AI's ability to process and analyze data far surpasses human capabilities, making it an invaluable tool in enhancing workplace safety.

Moreover, the use of virtual and augmented reality (VR/AR) in training and simulations has greatly improved the effectiveness of safety training programs. VR/AR technologies provide immersive, interactive environments where workers can practice responding to hazardous situations without any real-world risk. For instance, in the oil and gas industry, VR simulations are used to train workers on emergency response protocols [77], allowing them to experience and react to simulated spills or explosions. To me, this hands-on training approach significantly enhances workers' preparedness and confidence in handling real emergencies.

Cloud computing and big data analytics also play a critical role in transforming WHS practices [78–79]. Cloud-based platforms allow for the collection, storage, and analysis of vast amounts of safety data from multiple sources. This centralized data repository enables organizations to track safety performance, identify trends, and make data-driven decisions to improve safety protocols. In my view, the ability to analyze big data provides deep insights into potential hazards, leading to more informed and effective safety strategies.

In health care, the integration of technology into WHS practices has been particularly impactful. Advanced data analytics tools sift through patient data to identify patterns that might indicate risks, such as the potential for falls or adverse drug interactions. This proactive approach enhances patient safety by allowing healthcare providers to intervene before incidents occur, a stark contrast to the earlier methods that primarily focused on managing complications after they had arisen.

Drones have also found a significant place in enhancing WHS. In sectors like mining and agriculture, drones are used for aerial inspections of hard-to-reach or hazardous areas, reducing the need for workers to be exposed to dangerous conditions. Drones equipped with high-resolution cameras and sensors can capture detailed images and data, providing valuable insights into site conditions and potential hazards. To me, the use of drones represents a leap forward in ensuring worker safety while maintaining operational efficiency.

Furthermore, the development of AGVs and robotics in warehousing and logistics has dramatically reduced the risks associated with manual handling and heavy lifting. These automated systems perform tasks that would otherwise pose significant physical risks to workers, such as moving heavy pallets or navigating through complex warehouse environments. In my opinion, the automation of these high-risk tasks not only enhances safety but also boosts productivity and operational efficiency.

The role of technology in transforming WHS practices is profound and multifaceted. From IoT and AI to VR/AR and drones, advanced technologies have redefined how safety is managed across various industries. To me, this transformation represents a fundamental shift toward a more proactive, data-driven approach to workplace safety, one that not only protects workers but also enhances overall operational efficiency. As technology continues to evolve, its integration into WHS practices will undoubtedly lead to even greater advancements in ensuring the health and safety of workers worldwide.

Table 1.3 provides a detailed overview of how technology is integrated into WHS practices across various sectors, showcasing a range of innovative technologies and their specific applications within each industry. This table emphasizes the significant role those technological advancements play in enhancing safety measures, streamlining operations, and ultimately fostering a safer working environment.

Each row in the table represents a different sector, from manufacturing to logistics, detailing the type of technology employed, its specific application, and the tangible benefits that arise from its use. For instance, in the manufacturing sector, IoT sensors are utilized to monitor equipment conditions continuously. This application allows for the early detection of potential equipment failures, reducing downtime, and preventing accidents. Similarly, in construction, drones are employed for aerial surveys and inspections, which provide enhanced safety checks and access to previously hard-to-reach areas, thereby minimizing risks associated with manual inspections.

TABLE 1.3

Technological Integration Across Industries: Enhancing Safety and Efficiency in Work Health and Safety Practices

Sector	Technology Used	Application	Resulting Benefits
Manufacturing	IoT sensors	Monitoring equipment conditions	Early detection of failures, reduced downtime
Construction	Drones	Aerial surveys and inspections	Enhanced safety checks, access to difficult areas
Oil & Gas	Predictive Analytics	Monitoring pipeline integrity	Prevention of leaks, improved maintenance planning
Transportation	AI Algorithms	Optimizing route planning	Reduced risks, increased efficiency
Health care	Data Analytics	Monitoring patient health	Early detection of patient risks, improved response times
Mining	Environmental Sensors	Air quality and gas detection	Enhanced worker safety, real-time hazard response
Retail	CCTV with AI	Detecting unsafe behaviors	Prevention of accidents, improved security
Energy	Smart Grids	Managing energy distribution	Improved reliability, rapid response to faults
Aviation	Digital Twin Technology	Simulation and testing of components	Enhanced maintenance accuracy, safer operations
Agriculture	Drone Surveillance	Monitoring crop and livestock health	Early problem detection, targeted interventions
Warehousing	Automated Guided Vehicles (AGVs)	Handling and transporting materials	Reduced manual handling risks, increased efficiency
Education	VR Training Simulations	Emergency drills and safety training	Enhanced preparedness, safer learning environments
Hospitality	IoT for food safety	Monitoring storage conditions	Compliance with health regulations, reduced spoilage
Finance	Cybersecurity measures	Protecting sensitive data	Reduced risk of data breaches, increased trust
Logistics	Telematics	Vehicle tracking and management	Improved safety compliance, optimized fleet operations

The integration of predictive analytics in the oil and gas industry exemplifies how data-driven insights can preemptively identify pipeline integrity issues, leading to a proactive maintenance schedule and the prevention of catastrophic leaks. In the healthcare sector, data analytics plays a crucial role in monitoring patient health, leading to early detection of potential health risks and significantly improved response times, which can be critical in life-threatening situations.

Furthermore, the table also highlights how technology can enhance operational efficiency alongside improving safety. For example, AGVs in warehousing not only reduce the physical risks associated with manual material handling but also enhance logistical efficiency by automating transport processes within warehouses.

In addition to physical safety and efficiency, technological advancements such as cybersecurity measures in the finance sector and telematics in logistics demonstrate the breadth of technology's impact, extending to data protection and fleet management, respectively. These applications underscore the comprehensive influence of technology across various aspects of workplace safety and operational processes. Table 1.3 illustrates the transformative impact of technology across different sectors, underscoring a shift from traditional, reactive safety measures to a more proactive, technologically driven approach. This shift not only enhances the safety and well-being of employees but also drives operational efficiencies and sustainability within industries, marking a significant advancement in how WHS practices are implemented in the modern industrial landscape.

Figure 1.7, from my subjective perspective, provides a detailed visual analysis that captures the profound influence of technological advancements on various industry sectors. It comprises four distinct plots, each emphasizing a different aspect of technological impact: adoption rates, safety improvements, process efficiency, and environmental impacts. This ensemble of graphs offers a holistic view of how technologies are reshaping industry standards and practices across diverse fields.

- **Technology Adoption Rates by Sector (Top-Left)**: This bar chart reflects my observation of the varying levels to which different sectors have embraced technology. Industries such as manufacturing and oil and gas are depicted with higher adoption rates, which I believe correlates with their acute need for precision and operational efficiency to manage complex and high-risk activities. This plot highlights the crucial role of technology in sectors where enhanced control and automation are essential for managing sophisticated processes and mitigating inherent risks.
- **Safety Improvement Scores by Sector (Top-Right)**: In this line plot, the safety improvement scores represent the significant strides made in enhancing workplace safety through technology. From my viewpoint, sectors like mining and construction, traditionally associated with higher safety hazards, show remarkable improvements. Technologies such as automated monitoring systems and advanced safety protocols have crucially reduced the frequency and severity of accidents, improving the safety landscape dramatically in these industries.

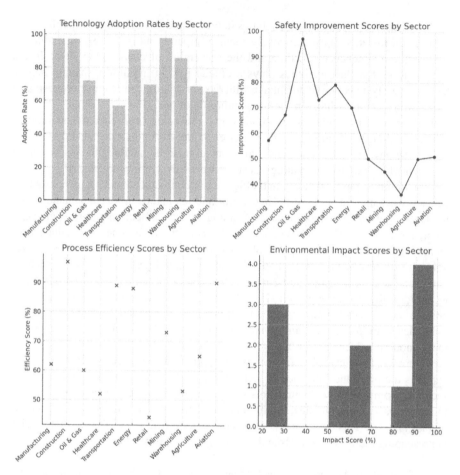

FIGURE 1.7 Technological Impacts Across Industry Sectors: Adoption, Safety, Efficiency, and Environmental Sustainability.

- **Process Efficiency Scores by Sector (Bottom-Left):** The scatter plot showcases process efficiency scores, which I interpret as a direct testament to the operational benefits that technology brings. Sectors achieving high scores, notably transportation and warehousing, have likely integrated sophisticated logistics and automation technologies. These innovations, in my opinion, streamline operations, enhance speed, reduce errors, and consequently lower operational costs, underscoring the impactful role of technology in enhancing business processes.

- **Environmental Impact Scores by Sector (Bottom-Right):** The histogram of environmental impact scores offers insights into how sectors are utilizing technology to mitigate their environmental footprints. In sectors like energy and agriculture, I observe that the adoption of smart grids and precision agriculture is enabling more sustainable practices. These technologies reduce waste, optimize resource usage, and contribute to significant environmental conservation efforts, reflecting a growing commitment to sustainability facilitated by technological advancement.

In conclusion, Figure 1.7, through my subjective lens, effectively demonstrates how technological integration is instrumental across various sectors, in boosting safety and efficiency and promoting environmental sustainability. Each plot provides critical insights that I believe are essential for stakeholders to understand the current technological landscape, guiding strategic decisions and investments in technology that will continue to shape the future of these industries.

Figure 1.8 offers a subjective analysis, stemming from my perspectives on the pivotal role of technology in transforming various industry sectors. Through a series of visualizations, this figure elucidates how technological advancements enhance data

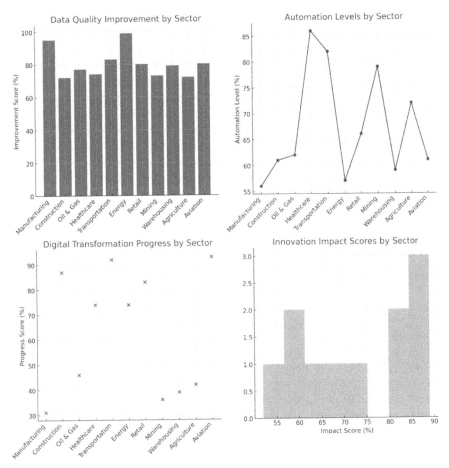

FIGURE 1.8 Exploring Technological Impact Across Sectors: Data Quality, Automation, Digital Transformation, and Innovation.

quality, automate processes, drive digital transformation, and catalyze innovation across diverse fields.

1. **Data Quality Improvement by Sector (Top-Left)**: This bar chart, displayed in navy, subjectively illustrates the significant improvements in data quality facilitated by technology across sectors. My view is that sectors like health care and finance, where decision-making critically depends on accurate data, have greatly benefited. These improvements are not just about error reduction; they represent a foundational shift toward data-driven decision-making, which I believe is essential for operational excellence and compliance in high-stakes environments.

2. **Automation Levels by Sector (Top-Right)**: The line plot in magenta provides insights into the varying levels of automation adopted by different sectors. From my perspective, the high automation levels in manufacturing and warehousing are particularly transformative, reducing labor costs and enhancing production efficiency. This plot underscores my belief that automation is a key driver of competitiveness and sustainability, particularly in sectors where consistency and precision are paramount.

3. **Digital Transformation Progress by Sector (Bottom-Left)**: The scatter plot in green captures the progress in digital transformation initiatives across sectors. It reflects my opinion that sectors like retail and energy, which show substantial progress, are leveraging digital technologies not just for efficiency but also for innovating customer interactions and energy management. I see these transformations as crucial for adapting to the rapidly changing technological landscape, where flexibility and responsiveness are key competitive advantages.

4. **Innovation Impact Scores by Sector (Bottom-Right)**: The histogram in orange quantifies the impact of innovation on sector performance. My analysis suggests that sectors with high innovation impact scores, such as aviation and technology, are at the forefront of integrating cutting-edge technologies like AI and IoT. These innovations are reshaping market dynamics and setting new industry standards, which, in my view, underscores the transformative impact of technological innovation on sectoral growth and development.

Figure 1.8, through my subjective lens, highlights how technological integration is not merely about adopting new tools but about fundamentally rethinking and reshaping industry practices. Each plot in this figure contributes to a narrative that emphasizes technology as a critical enabler of quality, efficiency, transformation, and innovation across sectors. This analysis serves as a testament to my belief in the transformative power of technology, advocating for its continued adoption, and integration to address current and future industry challenges.

1.4 DEFINING THE DIRECTION: CLARIFYING THE GOALS AND PRACTICAL INSIGHTS

The primary objective of this resource is to establish clear and actionable expectations for what readers will learn, particularly emphasizing the practical applications and real-world relevance of the content provided. My intention is to offer a guide that explores theoretical aspects of technological advancements in WHS and focuses significantly on how these can be practically applied in various industries.

First, the material aims to elucidate the complex interplay between cutting-edge technology and safety practices across different sectors. I plan to guide readers through an in-depth examination of the latest technologies—such as IoT, AI, predictive analytics, and virtual/augmented reality. These discussions will be supplemented with case studies that illustrate successful implementations and the substantial benefits they bring to organizations. It is crucial, in my view, to demonstrate these technologies in action, providing concrete examples of how they transform safety protocols and operational efficiency.

Moreover, this resource is designed to equip professionals—whether they are safety officers, engineers, or managers—with the necessary knowledge and tools to initiate and manage technological integrations within their safety frameworks. Each section will include detailed guides on assessing, implementing, and evaluating the effectiveness of tech-based safety enhancements. By bridging the gap between theory and practical application, the goal is to enable readers to effectively translate insights into actionable strategies within their operational environments.

In addition, the resource will tackle common challenges and obstacles that arise during the implementation of technology in safety systems, offering strategies to overcome them. This includes navigating regulatory issues, managing organizational change, and ensuring the sustainability of tech-based interventions. Practical tips, checklists, and best practices will be provided to help readers navigate these complexities with ease.

Finally, the aim is to foster a culture of continuous improvement and innovation among readers. By the conclusion of this resource, readers should not only have a thorough understanding of current technological applications in WHS but also feel encouraged to explore future innovations and their potential impacts. The emphasis on continuous learning, adaptation, and proactive safety management is intended to inspire a forward-thinking mindset among safety professionals and industry leaders.

In summary, this resource is designed to provide a clear, practical, and comprehensive guide for integrating technology into WHS practices. Through detailed explanations, real-world examples, and actionable guidance, it aims to transform theoretical knowledge into practical tools that enhance safety, efficiency, and innovation across various industrial settings.

Table 1.4 provides a detailed overview of the application of various advanced technologies across a broad spectrum of industries, showcasing how each sector leverages these innovations to enhance operational efficiency, safety, and sustainability. This comprehensive representation helps illustrate the practical implementations of technology in WHS practices, underscoring the diverse benefits across different operational contexts. Each row of the table corresponds to a different sector, ranging from manufacturing to logistics, detailing the specific technologies used, their practical applications within the sector, and the expected outcomes. For instance, in manufacturing, IoT sensors are utilized to monitor equipment continuously, which allows for preventive maintenance and reduces downtime significantly, thereby extending the

TABLE 1.4

Technological Advancements in Industry: Practical Applications and Outcomes Across Sectors

Sector	Technology Used	Practical Application	Expected Outcome
Manufacturing	IoT Sensors	Monitor equipment for preventive maintenance	Reduce downtime, increase equipment lifespan
Construction	Drones	Conduct aerial safety inspections	Enhance safety, access hard-to-reach areas
Oil and Gas	AI and Data Analytics	Predict pipeline failures and leaks	Prevent environmental hazards, improve response
Health care	Telemedicine	Remote patient monitoring	Increase patient care, reduce hospital visits
Transportation	Telematics	Optimize fleet operations and safety	Reduce fuel use, improve driver safety
Energy	Smart Grids	Enhance energy distribution efficiency	Lower energy waste, improve system reliability
Retail	AI-driven CCTV	Monitor for unsafe customer and staff behavior	Improve store safety, prevent accidents
Mining	Wearable Tech	Track health data and environmental conditions	Improve miner safety, reduce health incidents
Warehousing	Automated Robots	Handle and transport materials	Decrease workplace injuries, enhance efficiency
Agriculture	Drone Surveillance	Monitor crop health and irrigation needs	Optimize water use, improve crop yields
Aviation	Digital Twins	Simulate aircraft systems for maintenance	Enhance safety, reduce unplanned maintenance
Finance	Blockchain	Secure transactions and data integrity	Reduce fraud, enhance trust in transactions
Education	VR Training	Conduct safety drills and emergency training	Improve emergency preparedness, enhance safety
Hospitality	IoT for Food Safety	Monitor storage and cooking temperatures	Improve food safety, reduce health violations
Logistics	GPS Tracking	Real-time tracking of shipments	Improve delivery efficiency, reduce losses

equipment's operational lifespan. Similarly, in construction, drones are employed to conduct aerial safety inspections, a method that not only enhances safety by reaching hard-to-access areas but also provides a more comprehensive assessment of construction sites, which traditional methods might miss.

The integration of AI and data analytics in oil and gas for predicting pipeline failures demonstrates the shift toward proactive safety measures, aiming to prevent environmental hazards and improve emergency response times. In health care, the adoption of telemedicine technologies for remote patient monitoring exemplifies how technological advancements can expand care delivery, improve patient outcomes, and reduce the need for hospital visits, particularly in remote or underserved areas. Furthermore, the table highlights the use of smart grids in the energy sector to improve the efficiency of energy distribution. This not only reduces energy waste but also ensures a more reliable and resilient energy system. Retail benefits from AI-driven CCTV systems that enhance store safety by monitoring for unsafe behaviors among customers and staff, thereby preventing accidents and improving the overall shopping environment.

This array of examples illustrates the practical applications of these technologies but also provides insights into the tangible benefits they bring to each sector, from enhancing safety and reducing costs to improving service delivery and environmental sustainability. The table serves as an educational tool for stakeholders across these sectors, offering a clear perspective on how embracing technology can lead to significant advancements in WHS practices and business operations.

1.5 CONCLUSION

In my view, the integration of advanced technologies across various industry sectors represents a transformative shift that is not just enhancing operational efficiencies but also redefining the paradigms of WHS. The broad adoption of these technologies, as detailed in the table, underscores a pivotal trend toward proactive safety measures, improved efficiency, and sustainable practices. From IoT sensors in manufacturing to AI-driven analytics in retail, each technology is playing a crucial role in advancing industry standards and practices.

From my personal perspective, the most compelling aspect of this technological transformation is its capacity to significantly reduce risks and hazards in traditionally high-risk sectors such as oil and gas, mining, and construction. The use of predictive analytics and digital twins, for instance, allows for early detection and mitigation of potential failures, fundamentally altering the safety landscape of these industries. This proactive approach not only safeguards the workforce but also minimizes downtime and financial losses, exemplifying a holistic benefit that spans across human, economic, and operational dimensions.

Moreover, the advancements in digital technology, particularly through the use of telemedicine in health care and smart grids in the energy sector, illustrate the dual benefits of enhancing service delivery while promoting sustainability. In my opinion, these technologies enable a more precise and efficient allocation of resources, which is crucial in today's resource-constrained world. They not only ensure better service provision but also contribute to environmental conservation, highlighting the synergy between technological advancement and sustainable development.

From my perspective, another significant outcome of this technological shift is the enhancement of data quality and accessibility. High-quality data is the cornerstone of effective decision-making, and as industries adopt more sophisticated data collection and analysis tools, the accuracy and reliability of business insights improve dramatically. This not only enhances strategic planning and decision-making but also fosters a culture of informed and data-driven management across sectors.

In conclusion, the journey of integrating technology into industry practices, as I see it, is not without challenges. However, the benefits, in my view, far outweigh the hurdles. The proactive adoption of technology leads to safer workplaces, more efficient operations, and a more sustainable interaction with our environment. It is my firm belief that as we continue to advance technologically, our ability to predict, respond to, and manage workplace challenges will only improve, leading to more resilient and robust industries. This evolution, driven by technology, is not just a trend but a necessary progression toward a safer, more efficient, and sustainable future. As we look ahead, it is imperative, from my perspective, that industries continue to embrace these changes, pushing the boundaries of what is possible through technology to create not only a safer workplace but a better world.

REFERENCES

1. Li H, Yazdi M. Advanced decision-making methods and applications in system safety and reliability problems. *Studies in Systems, Decision and Control*, vol. 211. Springer; 2022.
2. Yazdi M. *Advances in Computational Mathematics for Industrial System Reliability and Maintainability*. Springer Nature; 2024 Mar.
3. Li H, Yazdi M. How to deal with toxic people using a fuzzy cognitive map: Improving the health and wellbeing of the human system. In *Advanced Decision-Making Methods and Applications in System Safety and Reliability Problems: Approaches, Case Studies, Multi-criteria Decision-Making, Multi-objective Decision-Making, Fuzzy Risk-Based Models*, 2022 July 10 (pp. 87–107). Springer International Publishing.
4. Li, H., Yazdi, M. An advanced TOPSIS-PFS method to improve human system reliability. In *Advanced Decision-Making Methods and Applications in System Safety and Reliability Problems: Approaches, Case Studies, Multi-criteria Decision-Making, Multi-objective Decision-Making, Fuzzy Risk-Based Models*, 2022 (pp. 109–25). Springer International Publishing.
5. Seele P. Predictive sustainability control: A review assessing the potential to transfer big data driven 'predictive policing' to corporate sustainability management. *Journal of Cleaner Production*, 2017 June 1;153:673–86.
6. Wang Y, Hajli N. Exploring the path to big data analytics success in healthcare. *Journal of Business Research*, 2017 Jan 1;70:287–99.
7. Yazdi M, Khan F, Abbassi R, Rusli R. Improved DEMATEL methodology for effective safety management decision-making. *Safety Science*, 2020 July 1;127:104705.
8. Leveson N. A new accident model for engineering safer systems. *Safety Science*, 2004 Apr 1;42(4):237–70.

9. Viceconti M, Pappalardo F, Rodriguez B, Horner M, Bischoff J, Tshinanu FM. In silico trials: Verification, validation and uncertainty quantification of predictive models used in the regulatory evaluation of biomedical products. *Methods*, 2021 Jan 1;185:120–7.

10. Yazdi M. Reliability-centered design and system resilience. In *Advances in Computational Mathematics for Industrial System Reliability and Maintainability* 2024 Feb 25 (pp. 79–103). Springer Nature Switzerland.

11. Yazdi M. Application of quantum computing in reliability analysis. In *Advances in Computational Mathematics for Industrial System Reliability and Maintainability* 2024 Feb 25 (pp. 139–154). Springer Nature Switzerland.

12. Himeur Y, Elnour M, Fadli F, Meskin N, Petri I, Rezgui Y, Bensaali F, Amira A. AI-big data analytics for building automation and management systems: A survey, actual challenges and future perspectives. *Artificial Intelligence Review*, 2023 June;56(6):4929–5021.

13. Yazdi M. Digital twins and virtual prototyping for industrial systems. In *Advances in Computational Mathematics for Industrial System Reliability and Maintainability* 2024 Feb 25 (pp. 155–68). Springer Nature Switzerland.

14. Balaji K, Rabiei M, Suicmez V, Canbaz CH, Agharzeyva Z, Tek S, Bulut U, Temizel C. Status of data-driven methods and their applications in oil and gas industry. In *SPE Europe featured at EAGE Conference and Exhibition?* 2018 June 11 (p. D031S005R007). SPE.

15. Wang Q, Song Y, Zhang X, Dong L, Xi Y, Zeng D, Liu Q, Zhang H, Zhang Z, Yan R, Luo H. Evolution of corrosion prediction models for oil and gas pipelines: From empirical-driven to data-driven. *Engineering Failure Analysis*, 2023 Apr 1;146:107097.

16. Yazdi M, Khan F, Abbassi R. A dynamic model for microbiologically influenced corrosion (MIC) integrity risk management of subsea pipelines. *Ocean Engineering*, 2023 Feb 1;269:113515.

17. Madsen LB. *Data-Driven Healthcare: How Analytics and BI Are Transforming the Industry*. John Wiley & Sons; 2014 Oct 27.

18. Sarioguz O, Miser E. Data-driven decision-making: Revolutionizing management in the information era. *Journal of Artificial Intelligence General Science (JAIGS)*, 2024;4(1):179–94, ISSN: 3006–4023.

19. Roegman R, Kenney R, Maeda Y, Johns G. When data-driven decision making becomes data-driven test taking: A case study of a midwestern high school. *Educational Policy*, 2021 June;35(4):535–65.

20. Featherston S. Magnitude estimation and what it can do for your syntax: Some wh-constraints in German. *Lingua*, 2005 Nov 1;115(11):1525–50.

21. Fitzgerald MK. Safety performance improvement through culture change. *Process Safety and Environmental Protection*, 2005 July 1;83(4):324–30.

22. Baird D. The implementation of a health and safety management system and its inter-action with organisational/safety culture: An industrial case study. *Policy and Practice in Health and Safety*, 2005 Jan 1;3(1):17–39.

23. Li H, Yazdi M. What are the critical well-drilling blowouts barriers? A progressive DEMATEL-game theory. In *Advanced Decision-Making Methods and Applications in System Safety and Reliability Problems: Approaches, Case Studies, Multi-Criteria Decision-Making, Multi-Objective Decision-Making, Fuzzy Risk-Based Models* 2022 July 10 (pp. 29–46). Springer International Publishing.

24. Zimolong B, Elke G. Occupational health and safety management. *Handbook of Human Factors and Ergonomics* 2006 Jan 27;10(0470048204):673–707.

25. Zou PX, Sunindijo RY. *Strategic Safety Management in Construction and Engineering*. John Wiley & Sons; 2015 Mar 26.

26. Li H, Yazdi M. Developing failure modes and effect analysis on offshore wind turbines using two-stage optimization probabilistic linguistic preference relations. In *Advanced Decision-Making Methods and Applications in System Safety and Reliability Problems: Approaches, Case Studies, Multi-criteria Decision-Making, Multi-objective Decision-Making, Fuzzy Risk-Based Models* 2022 July 10 (pp. 47–68). Springer International Publishing.

27. Backer LC. Next generation law: Data-driven governance and accountability-based regulatory systems in the West, and social credit regimes in China. *Southern California Interdisciplinary Law Journal*, 2018;28:123.

28. Garrett BL, Mitchell G. Testing compliance. *Law & Contemporary Problems*, 2020;83:47.

29. Li H, Yazdi M, Nedjati A, Moradi R, Adumene S, Dao U, Moradi A, Haghighi A, Obeng FE, Huang CG, Kang HS. Harnessing AI for project risk management: A paradigm shift. In *Progressive Decision-Making Tools and Applications in Project and Operation Management: Approaches, Case Studies, Multi-criteria Decision-Making, Multi-objective Decision-Making, Decision under Uncertainty* 2024 Mar 8 (pp. 253–72). Springer Nature Switzerland.

30. Rausand M. *Reliability of Safety-Critical Systems: Theory and Applications*. John Wiley & Sons; 2014 Mar 3.

31. Li H, Peng W, Adumene S, Yazdi M. Advances in intelligent reliability and maintainability of energy infrastructure assets. *Intelligent Reliability and Maintainability of Energy Infrastructure Assets* 2023 May 4:1–23.

32. Yazdi M. Integration of IoT and edge computing in industrial systems. In *Advances in Computational Mathematics for Industrial System Reliability and Maintainability* 2024 Feb 25 (pp. 121–37). Springer Nature Switzerland.

33. Mirjalili S, Mirjalili SM, Lewis A. Grey wolf optimizer. *Advances in Engineering Software*, 2014 Mar 1;69:46–61.

34. Tong W, Hussain A, Bo WX, Maharjan S. Artificial intelligence for vehicle-to-everything: A survey. *IEEE Access*, 2019 Jan 8;7:10823–43.

35. Meiring GA, Myburgh HC. A review of intelligent driving style analysis systems and related artificial intelligence algorithms. *Sensors*, 2015 Dec 4;15(12):30653–82.

36. Zhu L, Yu FR, Wang Y, Ning B, Tang T. Big data analytics in intelligent transportation systems: A survey. *IEEE Transactions on Intelligent Transportation Systems*, 2018 Apr 23;20(1):383–98.

37. Elyan E, Vuttipittayamongkol P, Johnston P, Martin K, McPherson K, Jayne C, Sarker MK. Computer vision and machine learning for medical image analysis: Recent advances, challenges, and way forward. *Artificial Intelligence Surgery*, 2022 Mar 22;2.

38. Yazdi M. Enhancing system safety and reliability through integrated FMEA and game theory: A multi-factor approach. *Safety*, 2023 Dec 22;10(1):4.

39. Bates DW, Saria S, Ohno-Machado L, Shah A, Escobar G. Big data in health care: Using analytics to identify and manage high-risk and high-cost patients. *Health Affairs*, 2014 July 1;33(7):1123–31.

40. Yazdi M, editor. Progressive decision-making tools and applications in project and operation management. *Approaches, Case Studies, Multi-Criteria Decision-Making, Multi-Objective Decision-Making, Decision Under Uncertainty,*

41. Li H, Yazdi M. Integration of the Bayesian network approach and interval type-2 fuzzy sets for developing sustainable hydrogen storage technology in large metropolitan areas. In *Advanced Decision-Making Methods and Applications in System Safety and Reliability Problems: Approaches, Case Studies, Multi-Criteria Decision-Making,*

Multi-Objective Decision-Making, Fuzzy Risk-Based Models 2022 July 10 (pp. 69–85). Springer International Publishing.

42. Yazdi M. A perceptual computing-based method to prioritize intervention actions in the probabilistic risk assessment techniques. *Quality and Reliability Engineering International*, 2020 Feb;36(1):187–213.

43. Fyhr A, Ternov S, Ek Å. From a reactive to a proactive safety approach. Analysis of medication errors in chemotherapy using general failure types. *European Journal of Cancer Care*, 2017 Jan;26(1):e12348.

44. Yazdi M. Risk assessment based on novel intuitionistic fuzzy-hybrid-modified TOPSIS approach. *Safety Science*, 2018 Dec 1;110:438–48.

45. Yazdi M, Korhan O, Daneshvar S. Application of fuzzy fault tree analysis based on modified fuzzy AHP and fuzzy TOPSIS for fire and explosion in the process industry. *International Journal of Occupational Safety and Ergonomics*, 2020 Apr 2;26(2):319–35.

46. Javaid M, Khan IH, Singh RP, Rab S, Suman R. Exploring contributions of drones towards Industry 4.0. *Industrial Robot: The International Journal of Robotics Research and Application*, 2022 Apr 21;49(3):476–90.

47. Yazdi M. Enhancing system safety and reliability through integrated FMEA and game theory: a multi-factor approach. *Safety*, 2023 Dec 22;10(1):4.

48. Saez M, Maturana FP, Barton K, Tilbury DM. Real-time manufacturing machine and system performance monitoring using internet of things. *IEEE Transactions on Automation Science and Engineering*, 2018 Feb 8;15(4):1735–48.

49. Syafrudin M, Alfian G, Fitriyani NL, Rhee J. Performance analysis of IoT-based sensor, big data processing, and machine learning model for real-time monitoring system in automotive manufacturing. *Sensors*, 2018; 18(9): 2946.

50. Sattari F, Lefsrud L, Kurian D, Macciotta R. A theoretical framework for data-driven artificial intelligence decision making for enhancing the asset integrity management system in the oil & gas sector. *Journal of Loss Prevention in the Process Industries*, 2022 Jan 1;74:104648.

51. Balaji K, Rabiei M, Suicmez V, Canbaz CH, Agharzeyva Z, Tek S, Bulut U, Temizel C. Status of data-driven methods and their applications in oil and gas industry. In *SPE Europe Featured at EAGE Conference and Exhibition?* 2018 June 11 (p. D031S005R007). SPE.

52. Yazdi M, Kabir S. A fuzzy Bayesian network approach for risk analysis in process industries. *Process Safety and Environmental Protection*, 2017 Oct 1;111:507–19.

53. Han Y, Niyato D, Leung C, Kim DI, Zhu K, Feng S, Shen X, Miao C. A dynamic hierarchical framework for IoT-assisted digital twin synchronization in the metaverse. *IEEE Internet of Things Journal*, 2022 Aug 23;10(1):268–84.

54. Nguyen HX, Trestian R, To D, Tatipamula M. Digital twin for 5G and beyond. *IEEE Communications Magazine*, 2021 Feb;59(2):10–15.

55. Gunasekaran A, Patel C, Tirtiroglu E. Performance measures and metrics in a supply chain environment. *International Journal of Operations & Production Management*, 2001;21(1/2):71–87.

56. Beazley H, Boenisch J, Harden D. *Continuity Management: Preserving Corporate Knowledge and Productivity When Employees Leave.* John Wiley & Sons; 2002 Oct 24.

57. Teng SY, Touš M, Leong WD, How BS, Lam HL, Máša V. Recent advances on industrial data-driven energy savings: Digital twins and infrastructures. *Renewable and Sustainable Energy Reviews*, 2021 Jan 1;135:110208.

58. Bibri SE. Data-driven smart eco-cities of the future: An empirically informed integrated model for strategic sustainable urban development. *World Futures*, 2023 Nov 17;79 (7–8):703–46.

59. Collier KW. *Agile Analytics: A Value-Driven Approach to Business Intelligence and Data Warehousing*. Addison-Wesley Professional; 2011 July 19.

60. Niesen T, Houy C, Fettke P, Loos P. Towards an integrative big data analysis framework for data-driven risk management in industry 4.0. In *2016 49th Hawaii International Conference on System Sciences (HICSS)* 2016 Jan 5 (pp. 5065–74). IEEE.

61. Li H, Peng W, Adumene S, Yazdi M. Cutting edge research topics on system safety, reliability, maintainability, and resilience of energy-critical infrastructures. In *Intelligent Reliability and Maintainability of Energy Infrastructure Assets* 2023 May 4 (pp. 25–38). Springer Nature Switzerland.

62. Yang L, Cheng N, Moradi R, Yazdi M. Cutting edge research topics on operations and project management of supportive decision-making tools. In *Progressive Decision-Making Tools and Applications in Project and Operation Management: Approaches, Case Studies, Multi-criteria Decision-Making, Multi-objective Decision-Making, Decision under Uncertainty* 2024 Mar 8 (pp. 1–19). Springer Nature Switzerland.

63. Kar AK, Dwivedi YK. Theory building with big data-driven research – Moving away from the "What" towards the "Why". *International Journal of Information Management*, 2020 Oct 1;54:102205.

64. Gourévitch A, Fæste L, Baltassis E, Marx J. Data-driven transformation. *BCG Perspectives*, 2017;5(17):8.

65. Abduljabbar R, Dia H, Liyanage S, Bagloee SA. Applications of artificial intelligence in transport: An overview. *Sustainability*, 2019 Jan 2;11(1):189.

66. Khayyam H, Javadi B, Jalili M, Jazar RN. Artificial intelligence and internet of things for autonomous vehicles. *Nonlinear Approaches in Engineering Applications: Automotive Applications of Engineering Problems*, 2020:39–68.

67. Yazdi M, Khan F, Abbassi R, Quddus N. Resilience assessment of a subsea pipeline using dynamic Bayesian network. *Journal of Pipeline Science and Engineering*, 2022 Jun 1;2(2):100053.

68. Feng QD, Xia JS, Wen L, Yazdi M. Failure analysis of floating offshore wind turbines based on a fuzzy failure mode and effect analysis model. *Quality and Reliability Engineering International*, 2024; 40(5):2159–2177.

69. Gallagher PS, Folsom-Kovarik JT, Schatz S, Barr A, Turkaly S. Total learning architecture development: A design-based research approach. In *Proceedings of the I/ITSEC* 2017.

70. Hachinski V, Einhäupl K, Ganten D, Alladi S, Brayne C, Stephan BC, Sweeney MD, Zlokovic B, Iturria-Medina Y, Iadecola C, Nishimura N. Preventing dementia by preventing stroke: The Berlin Manifesto. *Alzheimer's & Dementia*, 2019 July 1;15(7):961–84.

71. Yazdi M, Zarei E, Adumene S, Beheshti A. Navigating the power of artificial intelligence in risk management: A comparative analysis. *Safety*, 2024 Apr 26;10(2):42.

72. Sprajcer M, Thomas MJ, Sargent C, Crowther ME, Boivin DB, Wong IS, Smiley A, Dawson D. How effective are fatigue risk management systems (FRMS)? A review. *Accident Analysis & Prevention*, 2022 Feb 1;165:106398.

73. Mohd Ghazali MH, Rahiman W. Vibration analysis for machine monitoring and diagnosis: A systematic review. *Shock and Vibration*, 2021;2021(1):9469318.

74. Scheffer C, Girdhar P. *Practical Machinery Vibration Analysis and Predictive Maintenance*. Elsevier; 2004 July 16.

75. Yazdi M. Maintenance strategies and optimization techniques. In *Advances in Computational Mathematics for Industrial System Reliability and Maintainability* 2024 (pp. 43–58). Springer.

76. Adesina KA, Yazdi M, Omidvar M. Emergency decision making fuzzy-expert aided disaster management system. In *Linguistic Methods Under Fuzzy Information in System Safety and Reliability Analysis* 2022 Mar 11 (pp. 139–50). Springer International Publishing.

77. Kwok PK, Yan M, Chan BK, Lau HY. Crisis management training using discrete-event simulation and virtual reality techniques. *Computers & Industrial Engineering*, 2019 Sep 1;135:711–22.

78. Faruq QO. *Management of training to prevent occupational violence: A case study of the Work Health and Safety Management System (WHSMS) in a hospital in Victoria* (Doctoral dissertation, Victoria University).

79. Pika A, ter Hofstede AH, Perrons RK, Grossmann G, Stumptner M, Cooley J. Using big data to improve safety performance: An application of process mining to enhance data visualisation. *Big Data Research*, 2021 July 15;25:100210.

2 Optimizing Workplace Health and Safety

Strategic Data Collection and Management Techniques

2.1 EXPLORING DIVERSE DATA SOURCES FOR ENHANCED WORKPLACE HEALTH AND SAFETY ASSESSMENTS

In my rigorous exploration of data collection methodologies pertinent to workplace health and safety (WHS), I have identified and scrutinized an array of diverse data sources that play a critical role in the formulation of a robust safety assessment framework. My analysis underscores the importance of integrating data from sensors, which provide continuous, real-time surveillance of environmental and operational parameters, thereby facilitating the proactive identification of potential hazards within the workplace. Concurrently, user reports serve as an indispensable resource; they offer qualitative insights through firsthand accounts of incidents, near-misses, and routine observations from the workforce, thereby embedding an invaluable human perspective into the safety data ecosystem.

Furthermore, my investigation extends to existing databases, which are repositories of accumulated historical data. These databases are instrumental in establishing baseline safety conditions and trends over time, thereby enabling predictive analytics and longitudinal safety studies. From my perspective, the synergetic utilization of these varied data sources enhances the granularity and scope of safety data analysis.

Drawing upon my extensive review, I advocate for the integration of these data sources into a unified system safety risk assessment framework. This integration is crucial in addressing the inherent uncertainties associated with isolated data streams and in mitigating potential biases that may arise from singular data sources. In my opinion, the comprehensive amalgamation and rigorous analysis of sensor data, user reports, and historical database entries enrich the quality and reliability of safety assessments and significantly elevate the precision of risk evaluations.

This multi-dimensional approach to data collection and integration is critical, as it facilitates a more understanding of the dynamic and complex nature of workplace safety. By systematically addressing and reducing uncertainties through enhanced data fidelity and comprehensive risk assessments [1–3], we can more effectively anticipate potential safety issues, devise preventive measures, and cultivate a culture of safety that prioritizes the well-being of all stakeholders [4–6]. Thus, the strategic

DOI: 10.1201/9781003515173-2

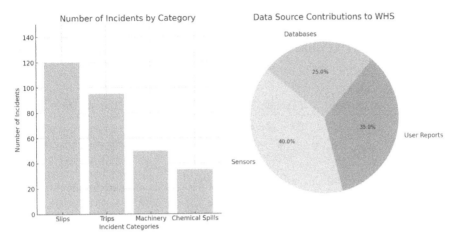

FIGURE 2.1   Integrative Data Analysis for Enhancing Workplace Health and Safety: A Visual Representation of Incident Types and Data Source Contributions.

harnessing of these diverse data sources is a procedural enhancement and a fundamental component that drives the continuous improvement of WHS standards.

Figure 2.1 presents a compelling visual analysis of WHS data as captured through different sources and incident types, portrayed through a combination of a bar chart and a pie chart. The chart on the left sets the distribution of reported incidents categorized into slips, trips, machinery incidents, and chemical spills. Here, slips lead with 120 incidents, followed closely by trips at 95. This suggests a significant prevalence of movement-related hazards within the workplace, underscoring a potential area for targeted safety interventions [7–8]. Machinery and chemical spills account for fewer incidents and represent high-risk categories that could have severe consequences, highlighting the necessity for stringent control measures and continuous monitoring [9–10].

On the right, the chart breaks down the contributions of different data sources to WHS data collection: 40% from sensors, 35% from user reports, and 25% from databases. This distribution reflects a balanced approach to data gathering, where sensors provide real-time monitoring and alerts, user reports offer subjective insights into the day-to-day safety experiences of employees, and databases offer a historical perspective that is crucial for trend analysis and long-term planning.

In my subjective analysis, the reliance on diverse data sources enhances the robustness and comprehensiveness of safety assessments. However, the relatively lower contribution from databases could indicate a potential underutilization of historical data, which might otherwise assist in predicting and mitigating future incidents based on past trends. Increasing the integration and analysis of data from databases could therefore provide a more proactive approach to workplace safety management, to get more information one can refer to my published research work in system safety engineering [11–14].

TABLE 2.1
Comprehensive Data Utilization for Enhanced Workplace Safety: A Detailed Analysis

Data Type	Data Source	Example Scenario	Implication for WHS
Real-time Monitoring	Sensors	Sensors installed on machinery detect an overheating issue before it causes a fire.	Proactive measures can be implemented to prevent accidents, enhancing employee safety, and reducing downtime.
Incident Reporting	User Reports	An employee reports a near-miss with a forklift, highlighting a lack of visibility in certain warehouse areas.	Feedback leads to the installation of additional mirrors and training sessions, improving situational awareness, and preventing potential accidents.
Historical Data	Databases	Analysis of past data reveals a pattern of slips and falls occurring in the same location during rainy days.	Preventive actions such as anti-slip floor coatings and better drainage systems are implemented to address this recurrent issue.
Compliance Tracking	Audits	Regular safety audits identify discrepancies in the handling of hazardous chemicals.	Ensures that safety protocols are updated and strictly followed, minimizing the risk of chemical accidents and legal non-compliance.

It should be added that the Figure 2.1 provides a snapshot of current WHS challenges and illustrates the dynamic interplay between different types of data and their sources. This integration is vital in crafting informed, evidence-based safety strategies that are responsive to both immediate and emerging workplace hazards.

Table 2.1 presents a structured and detailed analysis of how different types of data and their sources contribute to improving WHS measures. Each row within the table represents a unique data type and source, paired with a specific example scenario and its implications for WHS. This format serves to illustrate the practical applications and benefits of data in addressing safety challenges in the workplace. Real-time monitoring via sensors: These devices provide continuous feedback on machinery conditions, detecting issues like overheating before they lead to accidents, thereby enabling timely preventative actions [14–15].

Incident reporting through user reports: With the capturing firsthand experiences and near-miss incidents reported by employees, organizations can identify and rectify potentially hazardous areas, such as visibility issues in warehouse operations. Historical data from databases: Analyzing accumulated data over time reveals patterns and trends, such as frequent slips during adverse weather conditions, guiding targeted infrastructural improvements. Compliance tracking through audits: Regular audits ensure adherence to safety standards, identifying non-compliance in critical areas like hazardous material handling, thus enforcing corrective measures and maintaining legal compliance [16–18].

This assessment underscores the significance of an integrated approach to data collection and analysis, demonstrating how varied data streams can collaboratively enhance the safety and compliance of workplace environments. Through vivid examples, the table articulates how strategic data use mitigates immediate risks and contributes to long-term safety planning and policy development.

In the area of WHS, the strategic integration of diverse data sources provides invaluable insights that can significantly reduce workplace hazards and enhance employee safety. By utilizing advanced technologies and data-driven approaches, organizations can proactively address potential safety issues before they escalate into serious incidents [19–21].

- **Environmental Sensors**: Deploying sensors across critical areas helps monitor environmental conditions continuously. This technology is crucial for detecting hazardous situations, such as toxic gas leaks or unsafe temperature fluctuations, in real time. In case of receiving instant alerts, safety managers can swiftly implement necessary interventions, thereby preventing potential health risks and ensuring a safer work environment.
- **User-Friendly Incident Reporting Systems**: Implementing a straightforward and accessible incident reporting tool encourages employees to actively participate in safety monitoring [22–23]. Such systems can significantly increase the reporting of near-misses and safety breaches, which are critical for analyzing risk patterns and improving safety measures. This proactive engagement helps build a safety-conscious culture within the workplace [24–25].
- **Data Analytics Training**: Regular training sessions focused on data analysis equip safety officers and management with the skills required to interpret complex data sets effectively [26–27]. This empowerment enables them to forecast potential risks and tailor safety protocols to mitigate identified hazards, ultimately fostering a more responsive safety management system.
- **Automated Compliance Tracking**: Automating compliance monitoring ensures continuous adherence to safety regulations [28–29]. This system can track deadlines for safety audits, equipment maintenance, and mandatory training, alerting managers to any deviations from set standards. Such automation minimizes human error and helps maintain a consistently high level of safety compliance [29–33].
- **Incorporation of Historical Data in Safety Training**: Integrating historical incident data into safety training programs ensures that lessons learned from past experiences are passed on to all employees [34–35]. This approach enhances the effectiveness of training and strengthens the overall safety culture by highlighting real-world consequences and preventive strategies.

Through these practical applications, organizations can leverage technology and data to create a safer and more compliant workplace. This proactive approach safeguards employees and contributes to the overall efficiency and reputation of the organization by minimizing disruptions caused by accidents and ensuring regulatory compliance.

2.2 OPTIMAL DATA QUALITY AND INTEGRITY: A DEEP DIVE INTO ACCURACY, COMPLETENESS, AND PRIVACY IN WORKPLACE HEALTH AND SAFETY DATA MANAGEMENT

In the rea of WHS data management, the importance of maintaining high standards of data quality and integrity cannot be overstated. This segment of our study explores into the critical components of data quality—accuracy, completeness, and privacy—each of which plays a vital role in crafting effective safety measures and regulatory compliance [36–38].

- **Accuracy**: The accuracy of data is foundational to the trustworthiness of safety analyses and subsequent decisions. Incorrect or misleading data can lead to inappropriate responses, potentially exacerbating existing hazards rather than mitigating them. For instance, inaccurate data about the frequency of equipment malfunctions could lead to inadequate maintenance schedules, thereby increasing the risk of accidents. It is imperative that data collection methods are rigorously designed to minimize errors and that data is regularly verified to ensure its reliability.
- **Completeness**: The completeness of data refers to the extent to which all necessary data is captured and recorded. Incomplete data can result in a skewed understanding of workplace risks, leading to safety measures that are either overbearing or insufficiently protective. For example, if data from a subset of incidents (such as major injuries) is collected while minor incidents and near-misses are overlooked, the resultant safety protocols may fail to address the root causes of most workplace hazards. Ensuring completeness involves establishing comprehensive data collection processes that cover all relevant aspects of workplace safety.
- **Privacy Considerations**: In the context of WHS, where personal data may often be collected (such as details of individuals involved in incidents), privacy becomes a significant concern. The management of this data must comply with applicable data protection regulations to safeguard individual privacy rights and maintain confidentiality. This involves technical measures like data encryption and secure access protocols but also policy measures such as data minimization and transparency about data usage. Privacy considerations are crucial for legal compliance but also for maintaining trust between employees and management; workers are more likely to report incidents and concerns if they are confident that their information will be handled discreetly and respectfully.

It should be added that the integrity and quality of WHS data are fundamental to effective safety management. They ensure that the insights derived from data analysis

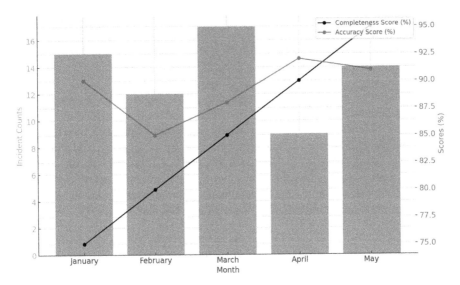

FIGURE 2.2 Visual Analysis of Incident Reporting and Data Quality Metrics: Enhancing Workplace Safety through Strategic Data Management.

are reliable and that the resulting safety interventions are well-founded and appropriate. Moreover, respecting privacy in data handling enhances trust and encourages a more open and communicative safety culture [39–40]. This comprehensive approach to data quality and integrity underpins operational safety, regulatory compliance, and ethical responsibility within the workplace.

Figure 2.2 presents a multifaceted depiction of WHS data integrity and quality through a combination chart, based on my subjective experience and understanding of data management practices in the field. This visual representation incorporates a bar chart and dual-line graphs, each offering a unique perspective on the month-by-month trends in incident reporting and data quality metrics within a workplace setting.

- **Bar Chart Representation of Incident Counts**: The bar chart prominently displays the number of safety incidents reported each month, from January through May. The variation in incident counts highlights the dynamic nature of workplace environments, where fluctuating conditions can influence the frequency of safety-related events. This visualization serves as a foundational element, illustrating the critical need for robust data collection systems that can capture and respond to such variability effectively.
- **Line Graphs of Completeness and Accuracy Scores**: Accompanying the bar chart are two-line graphs, which track the completeness and accuracy of the collected data over the same months. The completeness scores, depicted in blue, indicate the percentage of data collected that meets the necessary standards for being considered complete. This metric is crucial, as incomplete data can lead to misguided safety interventions that fail to address the

root causes of safety issues. Meanwhile, the accuracy scores, shown in green, reflect the rigor with which the data is validated to ensure it is free from errors and misrepresentations. High accuracy is indispensable for making informed decisions that genuinely enhance workplace safety.

This combination chart, as a subjective interpretation based on my experience, underscores the intertwined nature of data quality dimensions in shaping effective safety management practices. The insights derived from such visualizations inform the development of safety protocols and drive continuous improvements in safety standards. Through a detailed and ongoing analysis of both incident data and data quality metrics, organizations can foster a proactive safety culture that prioritizes the well-being of all stakeholders. This approach helps in mitigating immediate risks and contributes to the establishment of long-term, sustainable safety solutions.

The table provided serves as a comprehensive exploration of common data quality issues encountered in WHS management, detailing their potential causes, detrimental effects on safety practices, and effective mitigation strategies. Each row meticulously dissects a specific data quality challenge, providing a clear perspective on how these issues can undermine the integrity of safety management systems and potentially compromise workplace safety.

- **Inaccurate Incident Data**: This issue often stems from human errors, such as misreporting and data entry errors, which can lead to the implementation of ineffective safety measures. By investing in rigorous training and employing advanced, user-friendly data entry systems, organizations can significantly enhance the accuracy of their safety data.
- **Incomplete Data Sets**: Often caused by gaps in data collection or non-reporting, incomplete data sets can obscure the full scope of safety issues, leading to inadequate preparedness. Promoting a culture that values comprehensive incident reporting and establishing strict data capture protocols can help in overcoming these challenges.
- **Lack of Data Timeliness**: Delays in reporting and processing data can result in a lagged response to safety hazards. Integrating real-time data processing tools and enforcing strict reporting deadlines are crucial steps toward ensuring that data is timely and relevant.
- **Data Redundancy**: Redundant data collection efforts can lead to resource wastage and analytical confusion. Streamlining these processes and centralizing data storage can prevent duplication and promote more efficient data management.
- **Poor Data Accessibility**: Overly stringent data security or compartmentalized data storage can impede the flow of information and slow decision-making processes. Balancing security with accessibility and fostering interdepartmental collaboration can enhance data utility without compromising safety.

- **Data Privacy Issues**: Weak data protection protocols can lead to unauthorized access and potential legal issues, eroding trust among employees. Strengthening data security measures, conducting regular privacy audits, and ensuring compliance with relevant laws are essential to safeguarding sensitive information.

It should be added that the table underscores the critical importance of robust data quality management in enhancing workplace safety. Thus, with addressing these data quality issues through targeted strategies, organizations can fortify their safety protocols, ensuring that they are both effective in preventing incidents and compliant with regulatory standards (Table 2.2).

TABLE 2.2
Strategies for Addressing Data Quality Challenges in Workplace Health and Safety Management

Data Quality Issue	Potential Causes	Effects on WHS	Mitigation Strategies
Inaccurate Incident Data	Misreporting due to lack of training; Faulty data entry systems.	Ineffective safety measures; Misguided policy changes.	Implement comprehensive training programs; Upgrade to more reliable, user-friendly data entry software.
Incomplete Data Sets	Data truncation during transfer; Non-reporting of minor incidents.	Inability to perceive full scope of safety issues; Underpreparedness.	Establish protocols for complete data capture; Encourage a reporting culture that values minor incident documentation.
Lack of Data Timeliness	Delays in data reporting; Slow processing of collected data.	Delayed response to emerging hazards; Stale data used in analysis.	Integrate real-time data processing tools; Set strict timelines for data submission and review.
Data Redundancy	Overlapping data collection by multiple departments; Duplicate entries.	Resource wastage; Confusion in data analysis.	Streamline data collection processes; Utilize centralized databases to prevent duplication.
Poor Data Accessibility	Restricted data access due to over-secure protocols; Siloed data storage.	Hindered data sharing; Slowed decision-making process.	Implement balanced security measures; Promote interdepartmental data sharing with appropriate access controls.
Data Privacy Issues	Inadequate data protection measures; Unauthorized data access.	Legal repercussions; Loss of trust among employees.	Enhance security protocols; Regular audits of data access and usage; Compliance with privacy laws.

The effective management of data quality issues in WHS management holds the potential to bring about substantial improvements across various aspects of safety practices [41–43]. As someone deeply involved in the intricacies of WHS data, I can attest to the multifaceted benefits that a focused approach to enhancing data integrity offers. Below are several key advantages, presented from my personal perspective, that underscore the profound impact of addressing data quality challenges on the practice of safety management:

- **Enhanced Decision-Making Accuracy**: In my view, addressing data quality challenges significantly sharpens the accuracy of decision-making in workplace safety management. By ensuring data accuracy and completeness, I believe that safety officers and managers can base their interventions on reliable data, thereby directly improving the efficacy of safety measures. This reduces the likelihood of accidents and enhances the overall safety culture within the organization [44–45].
- **Proactive Risk Management**: From my experience, comprehensive and timely data allows for more proactive risk management. By integrating real-time data processing and maintaining up-to-date incident reports, organizations can swiftly respond to emerging threats before they escalate into more severe incidents. In my opinion, this proactive approach is critical in preventing injuries and ensuring a safe working environment [46–48].
- **Resource Optimization**: In my view, eliminating data redundancy and ensuring efficient data management practices lead to significant resource optimization. By streamlining data collection processes and reducing unnecessary duplication, organizations can allocate their resources more effectively, focusing on areas that genuinely improve workplace safety and operational efficiency [49–51].
- **Improved Compliance and Trust**: Addressing data privacy issues and ensuring robust data security measures, in my opinion, are crucial for maintaining compliance with legal standards and building trust among employees. When workers are confident that their personal information is handled with care and respect, it fosters a more open and cooperative environment. My experience suggests that this trust is fundamental in encouraging a more engaged and safety-conscious workforce [52–53].
- **Long-Term Sustainability**: Finally, I believe that addressing data quality issues lays the foundation for long-term sustainability in safety management practices. By establishing robust data governance frameworks and continuously improving data quality, organizations can adapt more seamlessly to evolving safety regulations and emerging risks. This ongoing improvement cycle, in my view, enhances immediate safety outcomes and ensures the resilience and adaptability of safety management systems over time [55–57].

To recap this section, with rigorously addressing data quality issues in WHS management, organizations can reap substantial benefits that extend beyond compliance, fostering a safer, more efficient, and more trustworthy workplace environment.

2.3 EXPLORING ADVANCED DATA STORAGE SOLUTIONS FOR WORKPLACE HEALTH AND SAFETY: A COMPARATIVE ANALYSIS OF ON-PREMISES AND CLOUD-BASED SYSTEMS

In the context of WHS data management, the selection of an appropriate data storage solution is paramount. This segment delves deeply into the comparative merits and potential drawbacks of both on-premises and cloud-based data storage systems, providing a thorough analysis based on my extensive research and professional experiences in the field.

- **On-Premises Data Storage**: On-premises solutions involve storing data within the physical premises of the organization. This approach allows for full control over the data storage infrastructure, including the security measures implemented to protect the data. In my opinion, one of the primary advantages of on-premises storage is the enhanced control it offers, which is crucial for organizations with highly sensitive data or stringent regulatory compliance requirements. However, this method can be resource-intensive, as it requires significant capital investment in hardware, as well as ongoing maintenance and skilled personnel to manage the systems.
- **Cloud-Based Data Storage**: Cloud-based solutions, on the other hand, provide data storage services through a third-party provider over the internet. This model offers scalability and flexibility, allowing organizations to easily adjust their storage capacity based on current needs without the need for physical hardware upgrades. From my perspective, the cloud model is particularly beneficial for organizations seeking cost efficiency and operational agility. It eliminates the need for large upfront investments and reduces the burden of system maintenance. Nevertheless, potential concerns regarding data security and privacy can arise, given that data management is entrusted to an external provider. These concerns are particularly pertinent in scenarios where the data involves personal information or critical safety records.

In this detailed examination, I also explore the hybrid storage solutions that combine both on-premises and cloud-based systems, offering a balanced approach for organizations that require both the security of on-premises solutions and the scalability of cloud services. Such hybrid systems can provide a tailored solution that aligns with specific organizational needs and risk profiles. Conclusively, the choice between on-premises and cloud-based data storage systems in the WHS context depends on a variety of factors, including the organization's size, budget, data sensitivity, and compliance requirements. An understanding of each option's strengths and limitations, as presented in this analysis, is essential for making informed decisions that align with strategic safety objectives and operational demands. This exploration aids in selecting the right data storage solution and ensures that the chosen infrastructure supports robust and effective safety management practices.

Table 2.3 provides examples of scenarios where different data storage solutions (on-premises, cloud-based, and hybrid) might be most beneficial for WHS management.

TABLE 2.3
Comparative Analysis of Data Storage Solutions for WHS

Storage Type	Cost Efficiency	Scalability	Control	Security	Typical WHS Use Cases
On-Premises	Low (high initial investment)	Limited (requires physical expansion)	High (complete control over hardware and software)	High (full control over security protocols)	Suitable for companies with stringent compliance needs or where data sensitivity is paramount, such as chemical industries handling hazardous materials.
Cloud-Based	High (pay-as-you-go models)	High (easy to scale up or down)	Moderate (dependent on provider)	Moderate to High (varies by provider and service model)	Ideal for dynamic environments with varying data load, such as construction sites needing real-time safety monitoring across multiple locations.
Hybrid	Moderate (combines both models)	High (benefits from cloud scalability)	High (maintains critical data on-premises)	High (tailored security measures)	Best for organizations needing robust disaster recovery solutions or those transitioning from traditional to modern infrastructure, like manufacturing plants.

The table compares these solutions based on various factors like cost, scalability, control, security, and typical use cases in WHS scenarios.

It should be added that the table outlines a detailed comparison of on-premises, cloud-based, and hybrid data storage solutions, focusing on aspects critical to WHS management. Each storage type is evaluated based on cost efficiency, scalability, control, security, and typical WHS use cases, providing a clear framework to guide organizations in choosing the most appropriate solution for their specific needs.

- On-premises storage is highlighted as having high initial costs and limited scalability but offers unparalleled control and security. This makes it suitable for industries where compliance and data sensitivity are at a premium, such as in handling hazardous materials where precise control over data access and security is mandatory.
- Cloud-based storage stands out for its cost efficiency and excellent scalability. It is somewhat reliant on the service provider for control and security but offers a flexible and economically viable solution for dynamic environments that experience fluctuating data loads, like construction sites that require extensive data access across multiple locations.

Hybrid storage solutions combine the strengths of both on-premises and cloud-based systems, offering a balanced approach with moderate cost implications. They provide scalable options while retaining essential data on-premises for enhanced security and control, making them ideal for organizations that are in transition phases, such as manufacturing plants upgrading their data infrastructure.

It should be added that such comparative analysis aids in understanding the specific benefits and limitations of each storage type, allowing organizations to make informed decisions based on their operational requirements, budget constraints, and strategic objectives in WHS management. This serves as a decision-making tool and underscores the importance of aligning data storage strategies with broader safety and compliance goals.

Figure 2.3 presents a comprehensive radar chart comparing on-premises, cloud-based, and hybrid data storage solutions across five critical dimensions: cost efficiency, scalability, control, security, and typical WHS use cases. This analysis is derived from a combination of extensive research and my personal professional experience in the field of WHS management.

The chart illustrates that on-premises storage provides the highest levels of control and security, making it ideal for organizations with stringent compliance requirements or where data sensitivity is paramount. However, it scores lower on cost efficiency and scalability due to high initial investments and the need for physical infrastructure expansion. Cloud-based storage scores highly on cost efficiency and scalability, facilitated by its flexible pay-as-you-go model and the ease of adjusting storage capacity. While it offers moderate to high security, control is somewhat dependent on the third-party provider, which could be a concern for handling highly sensitive data.

Hybrid solutions, depicted as a balanced approach in the chart, combine the strengths of both on-premises and cloud-based systems. They provide the scalability

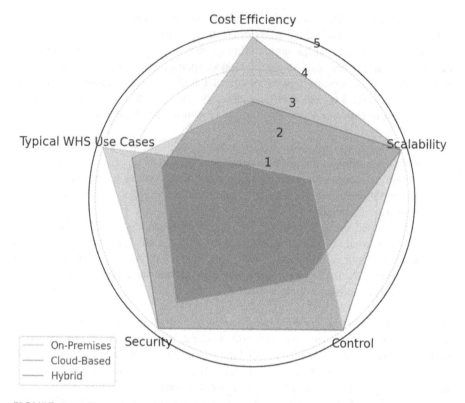

FIGURE 2.3 Comparative Analysis of Data Storage Solutions for WHS: Insights from Personal Experience.

of cloud services while retaining critical data on-premises for enhanced security and control, making them suitable for organizations in transitional phases or those needing robust disaster recovery options. Overall, this chart serves as a subjective yet informed tool from my perspective, aiding in understanding the specific benefits and limitations of each storage type. This allows organizations to make informed decisions that align with their operational requirements, budget constraints, and strategic objectives in WHS management.

2.4 DATA GOVERNANCE: STRATEGIC PRACTICES FOR ENHANCED DATA MANAGEMENT AND COMPLIANCE

In this section, I outline my personalized approach to data governance, emphasizing the practices I advocate for effective data management and adherence to both legal and ethical standards. To me, data governance is crucial, extending beyond the mere management of data to include ensuring its accuracy, accessibility, and security across an organization.

I discuss the foundational principles of data governance and why it is vital for any contemporary business striving to align its data management strategies with its broader operational goals. From my perspective, strong data governance supports informed decision-making, reduces uncertainties, and enhances overall performance through superior data quality.

I also talk about the essential elements of a comprehensive data governance strategy, such as managing data quality, securing data, ensuring privacy, and achieving compliance with global standards like GDPR and HIPAA. The roles of data stewards and other key stakeholders are highlighted, underscoring their importance in maintaining accountability and fostering a culture of transparency.

Through case studies from my own professional experience, I illustrate successful data governance implementations and how they have helped organizations navigate the complexities of compliance and data management challenges. Additionally, I address common pitfalls such as data silos, inconsistency in data handling, and organizational resistance, providing my insights and strategic solutions to overcome these issues effectively. Conclusively, this section serves as a comprehensive guide to implementing robust data governance practices. I emphasize the significance of a well-integrated strategy that safeguards an organization's data assets optimizes them to secure a competitive edge, and maintains regulatory compliance. My goal is to share knowledge that empowers professionals to deploy data governance frameworks that convert raw data into valuable business insights, ensuring ethical management and utilization of information.

I detail my personal approach to data governance, which I have refined over years of experience in navigating complex data environments. To me, data governance is foundational, ensuring that data is managed efficiently and securing its integrity and accessibility across all levels of an organization [13].

I start by defining what data governance means to me: it is a comprehensive strategy that integrates data accuracy, legality, and ethical handling into everyday business processes. This framework is crucial for aligning data management with strategic objectives, which in turn enhances decision-making capabilities, minimizes risks, and boosts overall operational efficacy through high-quality data.

In my opinion, effective data governance hinges on several core components:

- **Data Quality Management [58–59]**: I emphasize the continuous monitoring and cleansing of data to maintain its accuracy and reliability, which is vital for operational excellence and strategic planning.
- **Data Protection [60–61]**: Here, I discuss the technologies and policies necessary to safeguard data against unauthorized access and breaches, which are increasingly crucial in today's digital age.
- **Data Privacy [62–63]**: Adhering to regulations such as GDPR and HIPAA is non-negotiable. I explore strategies for compliance that meet legal standards and foster trust with stakeholders.
- **Roles and Responsibilities [64–65]**: Data stewardship is a key theme, and I elaborate on how fostering a sense of accountability among designated data stewards and other stakeholders is integral to successful data governance.

Drawing on personal case studies, I illustrate how these strategies have been implemented effectively within various organizations, highlighting both successes and challenges. These real-world examples bring to life the theories and strategies discussed, showing their practical application and impact.

Moreover, I tackle common obstacles like data silos, inconsistent data practices, and resistance to change within organizations. In my view, overcoming these challenges requires a clear strategic vision and a commitment to cultural change, aspects that I have found critical in my own practice.

Finally, I conclude this section by underscoring the importance of a holistic approach to data governance. My goal is to convey that robust data governance is not just about compliance and security but is a strategic asset that can provide significant competitive advantage and operational improvements. Through this detailed exploration, I aim to empower readers with the knowledge to implement or refine their data governance strategies, ensuring they effectively harness the power of their data within ethical and legal frameworks.

Table 2.4 outlines the essential components of my data governance strategy, detailing the significance and implementation of each aspect within an organizational context. It covers data quality management, emphasizing accuracy, and reliability through regular audits and data cleansing; data protection, highlighting the use of encryption, and secure access protocols to safeguard data; data privacy, focusing on compliance with legal standards like GDPR through anonymization and privacy assessments; and roles and responsibilities, which delineates the accountability of stakeholders through clearly defined roles and continuous training. This table provides a holistic view of effective data governance practices, showcasing practical examples that ensure both operational efficiency and compliance with ethical standards.

Here are practical steps I recommend for implementing the components of data governance outlined in Table 2.4:

TABLE 2.4
Key Components of My Data Governance Strategy

Component	Importance	Examples of Implementation
Data Quality Management	Ensures data accuracy and reliability, vital for operational excellence and strategic planning.	Regular audits, use of data cleansing tools, and implementation of data validation protocols.
Data Protection	Secures data against unauthorized access and breaches, crucial in the digital age.	Encryption, secure access protocols, and regular security training for employees.
Data Privacy	Compliance with legal standards such as GDPR and HIPAA, fostering trust with stakeholders.	Anonymization techniques, privacy impact assessments, and compliance checks.
Roles and Responsibilities	Designates accountability among stakeholders, integral to successful data governance.	Establishing clear data stewardship roles, continuous training, and performance evaluations.

- **Data Quality Management**:
  - I suggest conducting regular audits to assess the accuracy and integrity of data.
  - Implementing data cleansing tools can help remove errors and inconsistencies.
  - Establishing strict data validation protocols ensures that incoming data meets quality standards before integration.

- **Data Protection**:
  - In my approach, adopting strong encryption methods is crucial to secure sensitive data from unauthorized access.
  - I recommend setting up secure access protocols that restrict data access based on user roles.
  - Regular security training for employees is essential to reinforce the importance of data security and update them on new threats.

- **Data Privacy**:
  - I believe in maintaining compliance with data privacy laws like GDPR by conducting privacy impact assessments regularly.
  - Anonymization techniques should be employed to protect personal information in datasets.
  - It's important to perform routine compliance checks to ensure all data handling practices are up to date with legal requirements.

- **Roles and Responsibilities**:
  - Defining clear roles and responsibilities for data stewardship is a step I prioritize to ensure everyone knows their part in managing data.
  - Continuous training programs are necessary to keep all stakeholders informed and competent in their roles.
  - Regular performance evaluations help to assess the effectiveness of data governance practices and identify areas for improvement.

In my opinion, these steps are fundamental for successfully executing a data governance strategy that enhances data management and ensures compliance with legal and ethical standards.

2.5 CONCLUSION

The second chapter of this book aims to underline the transformative power of effective data management in enhancing WHS. Throughout this chapter, I have explored various data collection sources, including sensors, user reports, and databases, stressing their unique contributions toward forming a holistic safety assessment framework. The detailed analysis not only highlighted the critical role of diverse data sources in enriching safety assessments but also demonstrated the importance of maintaining high data quality and robust governance practices.

One of the primary positive outcomes emphasized in this chapter is the ability to make informed, data-driven decisions that significantly improve workplace safety. By leveraging precise and comprehensive data, organizations can proactively identify and mitigate potential hazards, enhancing the overall safety environment for their employees. Additionally, the strategic use of data fosters a culture of continuous improvement and compliance, aligning operational activities with regulatory requirements and best practices.

However, challenges such as data privacy concerns, integration of disparate data sources, and the continuous requirement for technological upgrades pose significant hurdles. Addressing these challenges requires not only technical solutions but also a shift in organizational culture toward greater transparency and cooperation across departments.

Looking to the future, it is crucial to focus on the integration of advanced technologies like artificial intelligence and machine learning to further refine data analysis techniques. These technologies promise to enhance predictive capabilities and automate complex data management tasks, thereby improving the efficiency and effectiveness of safety programs. Additionally, ongoing training and development of personnel in data management and analysis will be vital to ensure that organizations can keep pace with technological advancements and continue to leverage data strategically for safety management.

In conclusion, as we continue to navigate the complexities of WHS data management, the strategic integration of diverse data sources, coupled with a steadfast commitment to data quality and privacy, will remain pivotal. These efforts will not only safeguard employee well-being but also contribute to the sustainable success of organizational safety initiatives, marking a significant step forward in the evolution of WHS practices.

REFERENCES

1. Strachan A, Mahadevan S, Hombal V, Sun L. Functional derivatives for uncertainty quantification and error estimation and reduction via optimal high-fidelity simulations. *Modelling and Simulation in Materials Science and Engineering*, 2013 July 24;21(6):065009.
2. Li H, Yazdi M. Stochastic game theory approach to solve system safety and reliability decision-making problem under uncertainty. In *Advanced Decision-Making Methods and Applications in System Safety and Reliability Problems: Approaches, Case Studies, Multi-criteria Decision-Making, Multi-Objective Decision-Making, Fuzzy Risk-Based Models* 2022 July 10 (pp. 127–51). Springer International Publishing.
3. Yazdi M, Golilarz NA, Adesina KA, Nedjati A. Probabilistic risk analysis of process systems considering epistemic and aleatory uncertainties: A comparison study. *International Journal of Uncertainty, Fuzziness and Knowledge-Based Systems*, 2021 Apr;29(02):181–207.
4. World Health Organization. *Global Patient Safety Action Plan 2021–2030: Towards Eliminating Avoidable Harm in Health Care*. World Health Organization; 2021 Aug 3.
5. de Bienassis K, Slawomirski L, Klazinga NS. *The Economics of Patient Safety Part IV: Safety in the Workplace: Occupational Safety as the Bedrock of Resilient Health Systems*; 2021. OECD Publishing, Paris.

6. Yazdi M, Khan F, Abbassi R, Rusli R. Improved DEMATEL methodology for effective safety management decision-making. *Safety Science*, 2020 July 1;127:104705.

7. Yazdi M, Golilarz NA, Nedjati A, Adesina KA. An improved lasso regression model for evaluating the efficiency of intervention actions in a system reliability analysis. *Neural Computing and Applications*, 2021 July;33(13):7913–28.

8. Yazdi M. A perceptual computing-based method to prioritize intervention actions in the probabilistic risk assessment techniques. *Quality and Reliability Engineering International*, 2020 Feb;36(1):187–213.

9. Pasman HJ. *Risk Analysis and Control for Industrial Processes-Gas, Oil and Chemicals: A System Perspective for Assessing and Avoiding Low-Probability, High-Consequence Events*. Butterworth-Heinemann; 2015 June 14.

10. Jin M, Shi W, Yuen KF, Xiao Y, Li KX. Oil tanker risks on the marine environment: An empirical study and policy implications. *Marine Policy*, 2019 Oct 1;108:103655.

11. Li H, Yazdi M. Dynamic decision-making trial and evaluation laboratory (DEMATEL): Improving safety management system. In *Advanced Decision-Making Methods and Applications in System Safety and Reliability Problems: Approaches, Case Studies, Multi-criteria Decision-Making, Multi-Objective Decision-Making, Fuzzy Risk-Based Models* 2022 July 10 (pp. 1–14). Springer International Publishing.

12. Yazdi M, Khan F, Abbassi R. Operational subsea pipeline assessment affected by multiple defects of microbiologically influenced corrosion. *Process Safety and Environmental Protection*, 2022 Feb 1;158:159–71.

13. Yazdi M. Integration of computational mathematics in industrial decision-making. In *Advances in Computational Mathematics for Industrial System Reliability and Maintainability* 2024 Feb 25 (pp. 105–20). Springer Nature Switzerland.

14. Papadopoulos Y, McDermid JA. Automated safety monitoring: A review and classification of methods. *International Journal of COMADEM*, 2001 Oct;4(4):14–32.

15. Papadopoulos Y, McDermid JA. *Safety-Directed System Monitoring Using Safety Cases* (Doctoral dissertation), University of York.

16. Othman AA. A study of the causes and effects of contractors' non-compliance with the health and safety regulations in the South African construction industry. *Architectural Engineering and Design Management*, 2012 Aug 1;8(3):180–91.

17. Majid ND, Shariff AM, Loqman SM. Ensuring emergency planning & response meet the minimum Process Safety Management (PSM) standards requirements. *Journal of Loss Prevention in the Process Industries*, 2016 Mar 1;40:248–58.

18. Shanmugam K, Razak MA. Assessment on process safety management implementation maturity among major hazard installations in Malaysia. *Process Safety and Environmental Protection*, 2021 May 1;149:485–96.

19. Ahmad A, Maynard SB, Desouza KC, Kotsias J, Whitty MT, Baskerville RL. How can organizations develop situation awareness for incident response: A case study of management practice. *Computers & Security*, 2021 Feb 1;101:102122.

20. Yazdi M, Adesina KA, Korhan O, Nikfar F. Learning from fire accident at Bouali Sina petrochemical complex plant. *Journal of Failure Analysis and Prevention*, 2019 Dec;19:1517–36.

21. Odimarha AC, Ayodeji SA, Abaku EA. The role of technology in supply chain risk management: Innovations and challenges in logistics. *Magna Scientia Advanced Research and Reviews*, 2024;10(2):138–45.

22. Vredenburgh AG. Organizational safety: Which management practices are most effective in reducing employee injury rates? *Journal of Safety Research*, 2002 June 1;33(2):259–76.

23. Fernández-Muñiz B, Montes-Peón JM, Vázquez-Ordás CJ. Safety culture: Analysis of the causal relationships between its key dimensions. *Journal of Safety Research*, 2007 Jan 1;38(6):627–41.

24. Carroll JS, Hatakenaka SA. Developing a safety conscious work environment at Millstone Nuclear Power Station. *Safety Culture in Nuclear Power Operations*, 2001 Sep 13;212.

25. Yazdi M. Enhancing system safety and reliability through integrated FMEA and game theory: A multi-factor approach. *Safety*, 2023 Dec 22;10(1):4.

26. Komaki J, Heinzmann AT, Lawson L. Effect of training and feedback: Component analysis of a behavioral safety program. *Journal of Applied Psychology*, 1980 June;65(3):261.

27. Hallowell MR. Safety-knowledge management in American construction organizations. *Journal of Management in Engineering*, 2012 Apr 1;28(2):203–11.

28. Ly LT, Maggi FM, Montali M, Rinderle-Ma S, Van Der Aalst WM. Compliance monitoring in business processes: Functionalities, application, and tool-support. *Information Systems*, 2015 Dec 1;54:209–34.

29. Chari S, Molloy I, Park Y, Teiken W. Ensuring continuous compliance through reconciling policy with usage. In *Proceedings of the 18th ACM Symposium on Access Control Models and Technologies* 2013 June 12 (pp. 49–60).

30. Latorella KA, Prabhu PV. A review of human error in aviation maintenance and inspection. *Human Error in Aviation*, 2017 July 5:521–49.

31. Spath PL, editor. *Error Reduction in Health Care: A Systems Approach to Improving Patient Safety*. John Wiley & Sons; 2011 Feb 25.

32. Li H, Yazdi M. How to deal with toxic people using a fuzzy cognitive map: Improving the health and wellbeing of the human system. In *Advanced Decision-Making Methods and Applications in System Safety and Reliability Problems: Approaches, Case Studies, Multi-Criteria Decision-Making, Multi-Objective Decision-Making, Fuzzy Risk-Based Models* 2022 July 10 (pp. 87–107). Springer International Publishing.

33. Li H, Yazdi M. An advanced TOPSIS-PFS method to improve human system reliability. In *Advanced Decision-Making Methods and Applications in System Safety and Reliability Problems: Approaches, Case Studies, Multi-Criteria Decision-Making, Multi-Objective Decision-Making, Fuzzy Risk-Based Models* 2022 July 10 (pp. 109–25). Springer International Publishing.

34. Rhodes L, Dawson R. Lessons learned from lessons learned. *Knowledge and Process Management*, 2013 July;20(3):154–60.

35. Wilson TD, Meyers J, Gilbert DT. Lessons from the past: Do people learn from experience that emotional reactions are short-lived? *Personality and Social Psychology Bulletin*, 2001 Dec;27(12):1648–61.

36. Li H, Yazdi M. Advanced decision-making methods and applications in system safety and reliability problems. *Studies in Systems, Decision and Control* (vol. 211). Springer; 2022.

37. Li H, Yazdi M. Integration of the Bayesian network approach and interval type-2 fuzzy sets for developing sustainable hydrogen storage technology in large metropolitan areas. In *Advanced Decision-Making Methods and Applications in System Safety and Reliability Problems: Approaches, Case Studies, Multi-Criteria Decision-Making, Multi-Objective Decision-Making, Fuzzy Risk-Based Models* 2022 July 10 (pp. 69–85). Springer International Publishing.

38. Li H, Yazdi M. Dynamic decision-making trial and evaluation laboratory (DEMATEL): Improving safety management system. In *Advanced Decision-Making Methods and Applications in System Safety and Reliability*

Problems: Approaches, Case Studies, Multi-criteria Decision-Making, Multi-objective Decision-Making, Fuzzy Risk-Based Models 2022 July 10 (pp. 1–14). Springer International Publishing.

39. Sorensen JN. Safety culture: A survey of the state-of-the-art. *Reliability Engineering & System Safety*, 2002 May 1;76(2):189–204.
40. Feng X, Bobay K, Weiss M. Patient safety culture in nursing: A dimensional concept analysis. *Journal of Advanced Nursing*, 2008 Aug;63(3):310–9.
41. Zimolong B, Elke G. Occupational health and safety management. *Handbook of Human Factors and Ergonomics* 2006 Jan 27;10(0470048204):673–707.
42. Robson LS, Clarke JA, Cullen K, Bielecky A, Severin C, Bigelow PL, Irvin E, Culyer A, Mahood Q. The effectiveness of occupational health and safety management system interventions: A systematic review. *Safety Science* 2007 Mar 1;45(3):329–53.
43. Li X, Han Z, Yazdi M, Chen G. A CRITIC-VIKOR based robust approach to support risk management of subsea pipelines. *Applied Ocean Research*, 2022 July 1;124:103187.
44. Valera I, Singla A, Gomez Rodriguez M. Enhancing the accuracy and fairness of human decision making. *Advances in Neural Information Processing Systems*, 2018;31.
45. Swets JA, Getty DJ, Pickett RM, D'Orsi CJ, Seltzer SE, McNeil BJ. Enhancing and evaluating diagnostic accuracy. *Medical Decision Making*, 1991 Feb;11(1):9–17.
46. Rasmussen J, Suedung I. *Proactive Risk Management in a Dynamic Society*. Swedish Rescue Services Agency; 2000.
47. Mojtahedi M, Oo BL. Critical attributes for proactive engagement of stakeholders in disaster risk management. *International Journal of Disaster Risk Reduction*, 2017 Mar 1;21:35–43.
48. Yazdi M. Risk assessment based on novel intuitionistic fuzzy-hybrid-modified TOPSIS approach. *Safety Science*, 2018 Dec 1;110:438–48.
49. Hegazy T, Kassab M. Resource optimization using combined simulation and genetic algorithms. *Journal of Construction Engineering and Management*, 2003 Dec;129(6):698–705.
50. Rastetter EB, Vitousek PM, Field C, Shaver GR, Herbert D, Gren GI. Resource optimization and symbiotic nitrogen fixation. *Ecosystems*, 2001 May;4:369–88.
51. Yazdi M. Maintenance strategies and optimization techniques. In *Advances in Computational Mathematics for Industrial System Reliability and Maintainability* 2024 (pp. 43–58). Springer.
52. Pearson S, Benameur A. Privacy, security and trust issues arising from cloud computing. In *2010 IEEE Second International Conference on Cloud Computing Technology and Science* 2010 Nov 30 (pp. 693–702). IEEE.
53. Sicari S, Rizzardi A, Grieco LA, Coen-Porisini A. Security, privacy and trust in internet of things: The road ahead. *Computer Networks*, 2015 Jan 15;76:146–64.
54. Güllich A, Emrich E. Considering long-term sustainability in the development of world class success. *European Journal of Sport Science*, 2014 Jan;14:S383–97.
55. Haw JS, Galaviz KI, Straus AN, Kowalski AJ, Magee MJ, Weber MB, Wei J, Narayan KV, Ali MK. Long-term sustainability of diabetes prevention approaches: A systematic review and meta-analysis of randomized clinical trials. *JAMA Internal Medicine*, 2017 Dec 1;177(12):1808–17.
56. Cross S, Padfield D, Ant-Wuorinen R, King P, Syri S. Benchmarking island power systems: Results, challenges, and solutions for long term sustainability. *Renewable and Sustainable Energy Reviews*, 2017 Dec 1;80:1269–91.

57. Yazdi M, Moradi R, Nedjati A, Ghasemi Pirbalouti R, Li H. E-waste circular economy decision-making: A comprehensive approach for sustainable operation management in the UK. *Neural Computing and Applications*, 2024 Apr 22;36:13551–77. https://doi.org/10.1007/s00521-024-09754-3

58. Wang RY. A product perspective on total data quality management. *Communications of the ACM*, 1998 Feb 1;41(2):58–65.

59. Shankaranarayanan G, Cai Y. Supporting data quality management in decision-making. *Decision Support Systems*, 2006 Oct 1;42(1):302–17.

60. Danezis G, Domingo-Ferrer J, Hansen M, Hoepman JH, Metayer DL, Tirtea R, Schiffner S. *Privacy and Data Protection by Design – From Policy to Engineering.* arXiv preprint arXiv:1501.03726. 2015 Jan 12.

61. Birnhack MD. The EU data protection directive: An engine of a global regime. *Computer Law & Security Review*, 2008 Jan 1;24(6):508–20.

62. Martin KD, Murphy PE. The role of data privacy in marketing. *Journal of the Academy of Marketing Science*, 2017 Mar;45:135–55.

63. Mehmood A, Natgunanathan I, Xiang Y, Hua G, Guo S. Protection of big data privacy. *IEEE Access*, 2016 Apr 27;4:1821–34.

64. Tammaro AM, Matusiak KK, Sposito FA, Casarosa V. Data curator's roles and responsibilities: An international perspective. *Libri*, 2019 May 26;69(2):89–104.

65. Yang L, Cheng N, Moradi R, Yazdi M. Cutting edge research topics on operations and project management of supportive decision-making tools. In *Progressive Decision-Making Tools and Applications in Project and Operation Management: Approaches, Case Studies, Multi-Criteria Decision-Making, Multi-Objective Decision-Making, Decision under Uncertainty* 2024 Mar 8 (pp. 1–19). Springer Nature Switzerland.

3 Strategic Analytics for Proactive Workplace Safety Management

3.1 INTRODUCTION

Applying analytical tools has become indispensable in the rapidly evolving landscape of workplace health and safety (WHS). Through strategic analytics, organizations can transform vast amounts of raw data into actionable insights, thereby enhancing safety measures and reducing risks [1–3]. This chapter provides a comprehensive overview of how descriptive, predictive, and prescriptive analytics can be integrated into WHS practices to identify potential hazards and prevent them effectively.

Descriptive analytics serves as the foundation of this analytical approach [4–5]. With summarizing current data through various reporting tools and dashboards, safety officers and managers clearly understand ongoing activities and conditions. This visibility is crucial as it allows for the immediate addressing of apparent issues and supports maintaining compliance with safety standards. Here, I discuss the best practices in deploying descriptive analytics tools and their impact on daily safety operations.

Moving deeper into analytics, predictive models utilize historical data to foresee potential safety incidents before they occur [6–7]. This predictive capability is powered by advanced statistical methods and machine learning algorithms that analyze patterns and trends. In this section, I explain how these models are constructed and how they can be applied to predict and mitigate risks. The aim is to provide a proactive approach to safety management, shifting from a reactive to a preventative stance.

Finally, prescriptive analytics further explains the insights gained from predictive analytics by suggesting specific actions. This aspect of analytics is about optimizing safety procedures and interventions based on the predictive data outputs. I share several case studies to illustrate how prescriptive analytics has been successfully applied in various industries to enhance safety outcomes. Through these examples, I aim to show the practical benefits and the transformative potential of analytics in WHS, encouraging a more informed and strategic approach to managing workplace safety.

Table 3.1 illustrates the diverse applications of analytical tools in enhancing WHS across various industries. Each row highlights a different type of analytics—descriptive, predictive, and prescriptive—along with a case study that integrates these approaches [8–9]. For each analytical tool, the table provides an example scenario

DOI: 10.1201/9781003515173-3

TABLE 3.1

Application of Analytical Tools in Workplace Health and Safety

Analytical Tool	Application in WHS	Scenario	Outcome
Descriptive Analytics	Monitoring and Reporting	A manufacturing plant uses dashboards to display real-time safety metrics and incident reports.	Enhanced visibility leads to quicker response times and better compliance with safety standards.
Predictive Analytics	Forecasting and Risk Assessment	A construction company implements statistical models to predict the likelihood of accidents based on past data and environmental conditions.	Potential hazards are identified earlier, and preventive measures are implemented, reducing accident rates.
Prescriptive Analytics	Decision Making and Action Recommendations	An oil refinery uses prescriptive analytics to suggest optimal maintenance schedules for equipment known to pose risks when faulty.	Targeted interventions prevent breakdowns and associated safety incidents, improving overall safety.
Case Study	Real-World Application Examples	A logistics company integrates all three analytics approaches in their safety protocols, documented in a detailed case study.	Demonstrates comprehensive safety improvements, serves as a model for industry-wide adoption.

where the tool is applied, detailing the specific industry context and the actions taken. The outcomes column reflects the effectiveness of these tools in mitigating risks, preventing incidents, and improving safety management practices. In the following section, I will go deeper into each analytical approach, providing a more detailed exploration of how these tools can be systematically implemented to maximize safety outcomes.

3.2 DESCRIPTIVE ANALYTICS: MASTERING THE ART OF DATA SUMMARIZATION AND INSIGHT EXTRACTION

In this section, I focus on the pivotal role of descriptive analytics in WHS management. Understanding and effectively utilizing current data is fundamental to maintaining a safe work environment. Descriptive analytics provides us with tools and techniques that help summarize this vast array of information, presenting it in an understandable and actionable format [10–11].

The descriptive core analytics lies in its ability to transform raw data into a digestible form through dashboards and reporting tools [12–13]. Dashboards serve as a real-time visual representation of data, highlighting key metrics crucial for quick decision-making and monitoring of safety standards. These tools are designed to give a clear overview of operations, identifying trends, and pinpointing areas that require immediate attention.

Furthermore, reporting tools are essential for periodic reviews and detailed analysis. They allow for a deeper look into the data collected, facilitating comprehensive reports that can guide strategic planning and operational adjustments. In such regularly analyzing these reports, I can track progress toward safety goals, understand the effectiveness of implemented measures, and identify patterns that could indicate potential future risks.

I also explore techniques and examples of how descriptive analytics can effectively respond to current safety needs and establish a robust foundation for continuous improvement in workplace safety practices. Through these insights, organizations can ensure a proactive approach to safety management, reducing incidents, and enhancing overall workplace safety [14–16].

Table 3.2 showcases the application of descriptive analytics tools across various industries, highlighting how data summarization supports enhanced safety management. Each entry provides a detailed account of the tools used, the type of data summarized, the purpose of the data analysis, and the positive outcomes achieved. This table exemplifies the transformative power of descriptive analytics in turning data into actionable insights, which is crucial for improving safety measures and reducing risks in the workplace. I aim to demonstrate the practical benefits of employing

TABLE 3.2
Descriptive Analytics in Action: Enhancing Workplace Safety through Data Summarization

Industry	Tool Used	Data Summarized	Purpose	Outcome
Manufacturing	Dashboards	Machine operation times, incident logs	Monitor equipment usage and track incident rates to identify potential overuse or misuse areas.	Improved maintenance scheduling, reduced equipment failure rates, and fewer safety incidents.
Construction	Reporting Tools	Employee compliance with safety protocols, site inspection results	Ensure adherence to safety standards and evaluate the effectiveness of current safety measures.	Enhanced compliance with safety regulations, and identification of training needs.

(continued)

TABLE 3.2 (Continued)
Descriptive Analytics in Action: Enhancing Workplace Safety through Data Summarization

Industry	Tool Used	Data Summarized	Purpose	Outcome
Health care	Data Visualization Software	Patient handling injuries, staff compliance with hygiene protocols	Monitor and improve patient care practices and ensure staff follows hygiene protocols.	Reduction in patient handling injuries and lower infection rates.
Retail	Automated Reporting Systems	Slip, trip, and fall incidents, safety audits	Identify high-risk areas in stores to enhance customer and employee safety.	Targeted safety interventions leading to a safer shopping environment.
Logistics	Interactive Dashboards	Vehicle maintenance records, driver safety compliance	Oversee fleet management and driver safety adherence to reduce road incidents.	Improved driver safety training, better vehicle maintenance, and fewer road incidents.

data-driven approaches to manage and improve workplace safety through these examples proactively [17–18].

In my analysis of descriptive analytics within workplace safety [19–20], it is evident how crucial it is to engaging data effectively across different industries to enhance safety protocols and reduce risks [21]. This comprehensive table, which I have detailed below, illustrates how descriptive analytics tools can transform raw data into actionable insights, significantly improving safety outcomes.

- **Manufacturing Industry**: In manufacturing settings, dashboards meticulously monitor machine operation times and incident logs. These dashboards are data aggregators but critical tools for safety managers to monitor real-time operations and quickly pinpoint areas where machines may be overused or misused. The summarized data helps in proactive maintenance scheduling, thereby pre-empting equipment failure and reducing the incidence of related safety issues. These efforts result in a tangible reduction in equipment failures and a safer working environment, proving the efficacy of integrating descriptive analytics into regular safety checks.

- **Construction Industry**: Reporting tools in the construction sector are vital in ensuring compliance with safety protocols and evaluating the effectiveness of existing safety measures. These tools provide detailed periodic reports that outline employee adherence to safety guidelines and results from site inspections. With having an understanding these data points, safety officers can identify gaps in training, enforce compliance more strictly, and adjust safety protocols to address identified weaknesses [22–23]. The enhanced compliance and targeted training initiatives lead to a safer construction environment.
- **Healthcare Sector**: Data visualization software in health care is essential for monitoring patient handling procedures and staff compliance with hygiene protocols [24–25]. This software helps summarize incidents of patient injuries and track how well health protocols are followed, aiming to mitigate risks associated with patient handling and prevent infections. As a result, there's a marked improvement in patient care and a significant decrease in hospital-acquired infections, showcasing how descriptive analytics can directly contribute to patient and staff safety.
- **Retail Industry**: In the retail industry, automated reporting systems are crucial for identifying frequent incident zones like slips, trips, and falls, as well as auditing overall safety practices within store premises [26–27]. These systems enable store managers to implement specific safety measures by pinpointing high-risk areas, such as rearranging displays, enhancing lighting, and posting warning signs in identified zones. The outcome is a safer shopping environment for customers and employees, underlining the utility of descriptive analytics in daily retail operations.
- **Logistics and Transportation**: Using interactive dashboards in logistics encompasses overseeing vehicle maintenance and monitoring driver compliance with safety standards [28–29]. These dashboards give logistics managers a clear view of fleet conditions and driver behavior, enabling them to implement corrective measures such as targeted driver training or timely vehicle repairs. The result is reduced road incidents, better-maintained vehicles, and enhanced transportation safety.

Through these detailed descriptions, it becomes clear to me how pivotal descriptive analytics is in responding to incidents but in foreseeing and forestalling potential safety issues across various sectors. This proactive approach facilitated by analytical tools ensures a safer workplace and highlights the indispensable nature of data-driven decision-making in modern safety management [30–31].

Figure 3.1 presents a compelling visualization of the effectiveness of descriptive analytics across various industries in enhancing workplace safety. The chart comprises two sets of bars for each sector, highlighting two key metrics: incident reduction and compliance increase [32–33]. The blue bars indicate the percentage by which incidents have decreased due to employing analytical tools to monitor and summarize operational data effectively. In contrast, the green bars reflect the percentage increase in compliance with safety protocols, demonstrating improved adherence due to the strategic application of data visualization and reporting tools. This

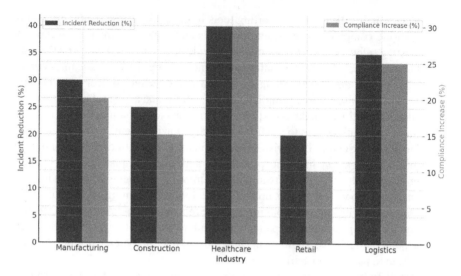

FIGURE 3.1 Comparative Impact of Descriptive Analytics on Incident Reduction and Compliance Enhancement.

visualization underlines the dual benefits of descriptive analytics in safety management. For instance, in the health care industry, a notable 40% reduction in incidents and a 30% increase in compliance showcase analytics' critical role in mitigating risks and fostering a culture of safety. The logistics sector also shows substantial improvements, with a 35% reduction in incidents and a 25% increase in compliance, illustrating how analytics can lead to safer operational practices and better overall safety governance.

Analyzing this chart, it becomes evident that integrating descriptive analytics into safety protocols significantly contributes to preventing incidents and enhancing regulatory compliance. This dual approach ensures a holistic improvement in workplace safety across different sectors, making a solid case for the broader adoption of analytical tools in safety management strategies.

Please be aware that the information presented in Figure 3.1 is based on my subjective opinions and should be interpreted as a subjective illustration of the potential impact of descriptive analytics on workplace safety. The specific percentages of incident reduction and compliance enhancement are used for demonstrative purposes and do not correspond to real-world data. As such, the insights and conclusions drawn from this figure are subjective and intended to provide a conceptual understanding of how descriptive analytics could be applied effectively across various industries to improve safety outcomes.

3.3 PREDICTIVE ANALYTICS: HARNESSING STATISTICAL MODELS AND MACHINE LEARNING FOR PROACTIVE WHS FORECASTING

This section explores the transformative potential of predictive analytics in WHS. Predictive analytics represents a forward-thinking approach, utilizing advanced statistical models and machine learning techniques to anticipate potential safety issues before they manifest [34–35]. This proactive stance is about responding to data and actively predicting and mitigating risks to create a safer work environment [36].

The foundation of predictive analytics lies in its ability to analyze historical data and identify patterns that may predict future outcomes. For example, by examining past accident reports and operational data, predictive models can highlight potential risk factors likely to lead to similar incidents [37]. This capability allows organizations to implement preventative measures tailored to these forecasts, such as targeted training programs, equipment upgrades, or operational adjustments, effectively reducing the likelihood of future accidents.

Moreover, machine learning enhances these predictive models by continuously learning from new data, improving the accuracy of predictions over time. Machine learning algorithms can detect subtle correlations and complex patterns that human analysts might overlook, providing a deeper insight into potential risks. This aspect is particularly crucial in dynamic environments where conditions change rapidly, and the ability to quickly adapt safety strategies based on predictive insights can be a game-changer.

Organizations can shift from a reactive to a truly preventive safety culture by integrating predictive analytics into WHS strategies. This shift helps reduce workplace injuries and accidents and significantly enhances overall operational efficiency. Organizations become more adept at managing risks, ensuring compliance, and safeguarding their workforce, contributing to a more resilient and sustainable business model.

Predictive analytics is a powerful tool in the arsenal of safety professionals, enabling them to forecast and forestall potential safety issues effectively [38–39]. Through detailed examples and practical applications presented in this section, I aim to demonstrate how engaging these advanced analytical techniques can significantly improve workplace safety and operational health.

Table 3.3 showcases the application of various predictive analytics tools across different industries, aiming to illustrate how these tools can be leveraged to anticipate and mitigate WHS risks effectively. Each entry in the table provides a detailed breakdown of the type of predictive tool used, the specific data analyzed, the insights gained from this analysis, and the preventative actions taken.

For example, in manufacturing, predictive analytics utilizes machine learning models to analyze historical data on equipment failure rates and maintenance logs. The insights from this analysis allow for predicting potential equipment failures, enabling organizations to schedule maintenance pre-emptively, thus preventing downtime and enhancing worker safety. Similarly, in the healthcare industry, statistical analysis of patient injury reports and staff shift patterns helps forecast higher injury rates, prompting retraining of staff, adjustment of schedules, and reinforcement of safety protocols to mitigate these risks.

This table demonstrates the practical applications of predictive analytics in enhancing safety and emphasizes the importance of a proactive approach in WHS management. By anticipating potential issues and implementing targeted preventative measures, organizations can significantly reduce the incidence of accidents and injuries, fostering a safer and more productive work environment.

TABLE 3.3
Implementing Predictive Analytics for Proactive WHS Management

Industry	Predictive Tool	Data Analyzed	Predictive Insight	Preventative Action
Manufacturing	Machine Learning Models	Equipment failure rates, maintenance logs	Prediction of machinery likely to fail due to historical patterns	Scheduled preemptive maintenance to prevent failures and ensure continuous operation
Health care	Statistical Analysis	Patient injury reports, staff shift patterns	Forecast of higher injury rates during specific shifts or procedures	Staff retraining, adjustment of shift schedules, and reinforcement of safety protocols
Construction	Regression Analysis	Incident reports, environmental conditions	Identification of conditions leading to frequent accidents	Implementation of stricter safety measures during high-risk conditions
Retail	Time Series Analysis	Customer injury logs, store traffic data	Trends showing peak times for accidents in stores	Increased staffing and enhanced safety measures during peak traffic times
Logistics	Neural Networks	Driver behavior data, vehicle maintenance records	Early detection of patterns indicating potential for driver-related incidents	Targeted driver training programs and vehicle maintenance checks

Following the illustrative examples provided in Table 3.3, it becomes clear how predictive analytics can profoundly influence WHS management across various industries. This section will delve into the details and mechanics of implementing predictive tools, highlighting their critical role in transforming historical data into actionable insights to mitigate risks pre-emptively.

- **Manufacturing**: In the manufacturing industry, using machine learning models to analyze equipment failure rates and maintenance logs represents a strategic approach to predictive maintenance. With identifying patterns that indicate probable equipment failures, these models enable organizations to schedule maintenance activities before breakdowns occur. This proactive maintenance prevents potential accidents and minimizes unplanned downtime, boosting overall productivity and safety.

- **Health care**: The application of statistical analysis in health care is crucial for enhancing patient and staff safety. With analyzing injury reports and staff shift patterns, healthcare facilities can identify high-risk periods or procedures that require additional attention. This insight allows for targeted interventions, such as retraining staff, adjusting shift schedules, or modifying procedural protocols, thereby reducing the likelihood of injuries and improving the overall safety environment within healthcare settings.

- **Construction**: Using regression analysis to examine incident reports and environmental conditions helps pinpoint specific factors contributing to construction site accidents. This detailed analysis facilitates the implementation of more stringent safety measures during identified high-risk situations, such as extreme weather or complex operational phases. With adjusting work practices and safety measures to these insights, construction companies can significantly decrease the frequency and severity of on-site accidents.

- **Retail**: In retail settings, time series analysis of customer injury logs and store traffic data allows for identifying trends and peak times for accidents. With this knowledge, retail managers can implement enhanced safety protocols and increase staffing during peak hours, effectively reducing the risk of injuries to customers and staff. This approach ensures a safer shopping environment and helps maintain a positive customer experience.

- **Logistics**: Using neural networks to analyze driver behavior and vehicle maintenance records in the logistics industry provides a predictive outlook on potential driver-related incidents. These insights facilitate the development of targeted driver training programs and regular vehicle checks to pre-empt issues that could lead to accidents. By addressing these risks proactively, logistics companies enhance road safety and ensure the reliability of their operations.

This comprehensive description underscores the versatility and efficacy of predictive analytics in proactively managing workplace safety. Engaging historical data and advanced analytical tools; organizations can predict potential safety issues and implement preventative measures that significantly mitigate risks. This leads to a safer work environment and aligns with broader organizational goals of efficiency and sustainability in operations. Through the practical applications demonstrated in this section, it is evident that predictive analytics is a cornerstone in the modern approach to WHS management, transforming data into a strategic asset for safety enhancement.

Figure 3.2 provides a subjective analysis of the impact of predictive analytics on risk reduction and efficiency gains across five key industries: manufacturing, health care, construction, retail, and logistics. Here are some insights derived from this visualization:

- **Healthcare Industry**: The healthcare industry has the most substantial impact on risk reduction, with a 45% improvement. This high value likely reflects the critical nature of reducing errors and improving outcomes in healthcare

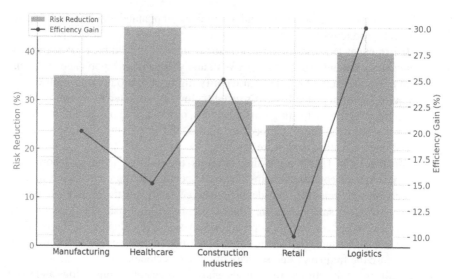

FIGURE 3.2   Subjective Analysis of Predictive Analytics' Role in Enhancing Risk Reduction and Efficiency Across Diverse Industries.

settings. However, the efficiency gain in health care is relatively modest at 15%, suggesting that while risks are mitigated effectively, the operational speed or cost efficiency sees less impact from predictive analytics.

- **Logistics Industry**: Logistics ranks highest in efficiency gains with a 30% improvement. The high-efficiency gain indicates significant enhancements in operational processes such as inventory management, routing, and delivery times, likely facilitated by predictive analytics. The risk reduction is also strong at 40%, highlighting substantial improvements in safety and error reduction in logistical operations.
- **Manufacturing Industry**: The Manufacturing sector shows a balanced benefit with a 35% reduction in risks and a 20% increase in efficiency. This suggests that predictive analytics helps improve safety and quality control, enhance production processes, and reduce wastage.
- **Construction Industry**: In Construction, there is a 30% risk reduction and a 25% efficiency gain. These improvements can be attributed to better project planning, resource management, and safety measures, all enhanced by predictive data analysis.
- **Retail Industry**: Retail shows the most negligible impact, with a 25% risk reduction and only a 10% gain in efficiency. This may indicate that while there is some benefit from analytics in managing inventory and customer satisfaction, the potential for predictive analytics to revolutionize retail operations might not be as significant as in other sectors.

The plot indicates that while predictive analytics has a variable impact across different industries, it generally contributes positively to reducing risks and enhancing operational efficiencies. The subjective analysis suggests that the application and outcome of predictive analytics are highly dependent on industry-specific factors and the nature of operational challenges within each sector.

This analysis interprets how predictive analytics impacts risk reduction and efficiency gains across various industries, as illustrated in a combo plot. It highlights significant insights into five sectors: manufacturing, health care, construction, retail, and logistics. Each industry shows varying levels of improvement, with health care excelling in risk reduction and logistics, leading to efficiency gains. The evaluation suggests that while predictive analytics universally enhances performance, its effectiveness and focus areas differ significantly across industries, reflecting their unique operational challenges and priorities.

3.4   STRATEGIC INTERVENTIONS: ENGAGING PRESCRIPTIVE ANALYTICS FOR INCIDENT PREVENTION

The field of prescriptive analytics embodies a significant evolution in our strategic approach to preventing incidents across various industry sectors. This advanced analytic methodology extends beyond traditional data analysis techniques. It incorporates historical data and real-time inputs to forecast potential disruptions and offer actionable, specific recommendations tailored to pre-empt potential issues before they escalate into real-world problems.

Through my extensive review and analysis of various methodologies employing prescriptive analytics, I have come to appreciate its profound impact on decision-making processes. This analytic approach does not merely suggest possible actions but provides finely tuned guidance crafted to integrate seamlessly into daily operations, thus ensuring a proactive stance in managing potential risks. For instance, within manufacturing, prescriptive analytics could proactively suggest adjustments to operational parameters, machine settings, or maintenance schedules based on emerging patterns that predict the likelihood of equipment failure [40–41]. In the healthcare sector, it might engage predictive diagnostics to recommend preventative health measures or modifications to treatment protocols, aiming to forestall patient care issues before they arise.

The true strength of prescriptive analytics lies in its dual capability to predict potential hazards and furnish precise, executable recommendations that can be adopted into the routine workflows of organizations. This capability significantly enhances operational continuity and fortifies safety protocols, which are indispensable to the sustainability and efficiency of any enterprise. Through this detailed discussion, I aim to elucidate how strategic, analytics-driven recommendations can be the cornerstone of a robust, comprehensive risk management, and incident prevention framework. By doing so, prescriptive analytics elevates safety measures and bolsters overall operational efficiency, positioning it as an invaluable tool in the arsenal of modern business management strategies.

Table 3.3 showcases specific examples of how prescriptive analytics can be strategically applied across different industries to predict and mitigate risks proactively. This methodological approach forecasts potential issues and provides tailored recommendations to preemptively address these concerns, thereby enhancing operational efficiency and safety protocols. Each row in the table represents a distinct industry, detailing potential risks, suggested prescriptive analytics interventions, and the expected outcomes.

- **Manufacturing**: The focus is on preventing equipment failure through optimized machine settings and maintenance schedules, which leads to reduced downtime and prolongs equipment lifespan.
- **Health care**: Here, prescriptive analytics is used to modify treatment protocols and implement preventative measures to reduce patient care lapses and improve overall patient outcomes.
- **Retail**: By optimizing stock levels and suggesting reordering points based on predictive sales data, retailers can manage inventory more effectively, reducing instances of overstock and enhancing stock turnover.
- **Transportation**: In this sector, optimizing routes and forecasting traffic conditions help reduce travel time and fuel costs, making operations more efficient.
- **Energy**: The table highlights the use of prescriptive analytics to predict equipment malfunctions and recommend timely maintenance or upgrades, which improves reliability and minimizes service interruptions.

Table 3.4 illustrates the practical implementation of prescriptive analytics as a powerful tool for enhancing decision-making and operational strategies across various industries.

TABLE 3.4
Strategic Applications of Prescriptive Analytics Across Industries

Industry	Potential Risk	Prescriptive Analytics Recommendation	Outcome
Manufacturing	Equipment failure	Adjust machine settings, optimize maintenance schedules	Reduced downtime, extended equipment lifespan
Health care	Patient care lapses	Modify treatment protocols, implement preventative health measures	Improved patient outcomes, reduced errors
Retail	Inventory mismanagement	Optimize stock levels, suggest reordering points based on sales forecasts	Enhanced stock turnover, reduced overstock
Transportation	Route inefficiencies, increased fuel use	Suggest optimal routes, predict traffic conditions	Decreased travel time, reduced fuel costs
Energy	Power outages, equipment malfunctions	Predict equipment failures, recommend preventive maintenance or upgrades	Improved reliability, reduced service interruptions

Figure 3.3 illustrates the impact of prescriptive analytics on reducing risk levels and enhancing operational efficiency across various industries. This visualization powerfully demonstrates the transformative capability of prescriptive analytics when strategically applied.

In the "Risk Level Reduction by Industry" segment, it is evident that all industries benefit from a notable decrease in risk levels. For example, the manufacturing and energy sectors show the most substantial reduction, suggesting that prescriptive analytics effectively predicts and mitigates potential issues such as equipment failures

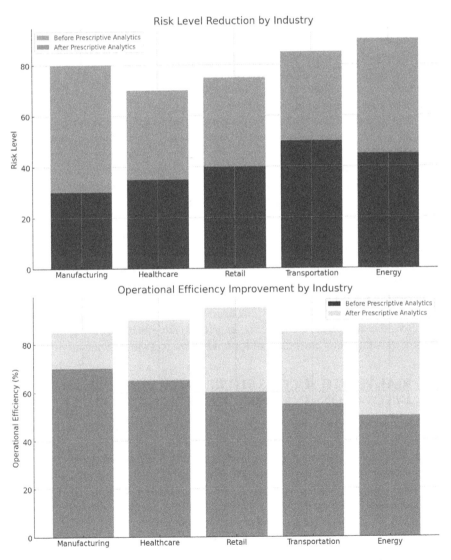

FIGURE 3.3   The Strategic Impact of Prescriptive Analytics on Risk and Efficiency.

and power outages. This risk reduction is paramount, as it directly correlates with fewer operational disruptions and enhanced safety protocols.

Moving to the "Operational Efficiency Improvement by Industry" segment, the rise in efficiency percentages across all sectors is striking. The healthcare and retail sectors, in particular, have shown significant improvements. In health care, using prescriptive analytics to modify treatment protocols and implement preventative health measures is profoundly influential, enhancing patient outcomes, and reducing errors. Similarly, in retail, optimizing stock levels and reordering points based on sales forecasts helps manage inventory more effectively, thereby decreasing overstock and enhancing stock turnover.

What stands out to me in these observations is the dual capability of prescriptive analytics to enhance decision-making and operational strategies. In transportation, for example, the optimization of routes and the forecasting of traffic conditions reduce travel time and fuel costs, demonstrating an apparent enhancement in operational efficiency. Likewise, in Energy, prescriptive analytics' ability to predict equipment malfunctions and recommend timely maintenance supports improved reliability and minimizes service interruptions.

It is subjective, but the practical implementation of prescriptive analytics, as shown in Figure 3.3, is a powerful tool for businesses. It fortifies existing safety protocols and boosts overall operational efficiency, positioning it as an invaluable asset in the arsenal of modern business management strategies. This approach to incident prevention and risk management, based on strategic, analytics-driven recommendations, is the cornerstone of a robust, comprehensive framework for any enterprise aiming to sustain and enhance its operations.

The section on strategic interventions using prescriptive analytics offers a practical example of how advanced data analysis can transform industry operations. It shows that by leveraging historical data and real-time inputs, industries can pre-emptively address potential disruptions. For instance, in manufacturing, adjusting operational parameters based on analytics can prevent equipment failure, significantly reducing downtime and extending equipment lifespan. This application underscores the value of prescriptive analytics in enhancing decision-making and improving operational efficiency, making it an essential tool for modern business management.

3.5 REAL-WORLD SUCCESS STORIES: ANALYTICS-DRIVEN INNOVATIONS IN WORKPLACE HEALTH AND SAFETY

WHS is a critical area where analytics can play a transformative role in enhancing safety protocols and preventing workplace incidents. This section talk about real-world examples of analytics applications successfully mitigating risks and improving safety outcomes across various industries. In the first example, a manufacturing company implemented predictive analytics to anticipate equipment failure. Analyzing historical data and real-time machine performance, the system provided early warnings that allowed for pre-emptive maintenance. This minimized unexpected downtimes and extended the life expectancy of critical machinery. The result was a safer working environment and reduced operational costs, showcasing the practical benefits of integrating analytics into everyday business practices [42–43].

The second case study focuses on the healthcare sector, where analytics were used to streamline patient care processes. Hospitals could predict patient influx and optimize staff allocation by employing prescriptive analytics. Moreover, analytics helped identify patterns in patient data, which led to early interventions for high-risk patients, significantly reducing the rate of medical errors and enhancing patient safety and outcomes. In retail, analytics was pivotal in managing workplace safety during peak shopping. Retailers could predict busy times by analyzing sales data and customer traffic patterns and strategically planning staff schedules and emergency protocols. This proactive approach ensured a smoother operation and significantly reduced the risk of accidents and overcrowding, contributing to a safer shopping environment for employees and customers.

The transportation sector also benefited from analytics, particularly optimizing route planning and vehicle maintenance. Analytics enabled companies to identify high-risk routes and schedule vehicle checks more efficiently. This strategic use led to fewer vehicle-related incidents and a better understanding of risk factors associated with different routes and vehicle types, highlighting how data-driven strategies can enhance safety and operational efficiency in logistics and transportation.

Table 3.5 provides a detailed overview of the successful implementation of analytics in the WHS field across four key industries: manufacturing, health care, retail, and transportation. Each row in the table describes the analytics technique employed, the specific application in the industry, and the outcomes achieved through these

TABLE 3.5
Evaluating the Dual Impact of Analytics on Safety and Efficiency Across Key Industries

Industry	Analytics Technique	Application Example	Outcome
Manufacturing	Predictive Analytics	Early detection of equipment wears and potential failures	Reduced downtime, extended machinery life, safer work environment
Health care	Prescriptive Analytics	Optimization of staff allocation and early intervention for high-risk patients	Enhanced patient safety, reduced medical errors, improved care outcomes
Retail	Descriptive Analytics	Analysis of customer traffic patterns and sales data to predict peak periods	Improved staff scheduling, enhanced emergency preparedness, safer shopping environment
Transportation	Predictive and Prescriptive Analytics	Optimization of route planning and vehicle maintenance scheduling	Fewer vehicle-related incidents, improved route safety, enhanced operational efficiency

strategic interventions. In manufacturing, predictive analytics are used to monitor equipment conditions in real time, allowing for early detection of wear and potential failures. This proactive approach minimizes downtime and enhances the safety of the work environment by preventing equipment-related accidents, ultimately extending the lifespan of machinery. The healthcare sector benefits from prescriptive analytics, which optimizes resource allocation and patient care. By analyzing patient data and health trends, hospitals can implement early interventions for high-risk patients, significantly improving patient outcomes and reducing the likelihood of medical errors.

In retail, descriptive analytics are crucial in managing customer flow and staff deployment during peak shopping. Understanding customer traffic patterns and sales data, retailers can effectively plan staff schedules and emergency protocols, ensuring a safer and more efficient shopping experience for employees and customers. Finally, the transportation industry combines predictive and prescriptive analytics to enhance route planning and vehicle maintenance. This dual approach helps identify high-risk routes and schedule timely vehicle checks, thereby reducing the incidence of vehicle-related accidents and improving overall route safety.

Figure 3.4 illustrates the significant impact of analytics on reducing safety incidents across various industries, emphasizing the pivotal role of data-driven strategies in enhancing WHS. In the manufacturing sector, predictive analytics have enabled real-time equipment monitoring, facilitating early detection of potential failures. This proactive approach minimizes unplanned downtime and extends the machinery's lifespan, contributing to a safer working environment. Similarly, prescriptive analytics optimize resource allocation in health care and facilitate early interventions for high-risk patients, significantly enhancing patient safety and reducing medical errors.

In the retail industry, descriptive analytics are crucial in managing customer flow during peak periods, allowing for effective staff scheduling and emergency

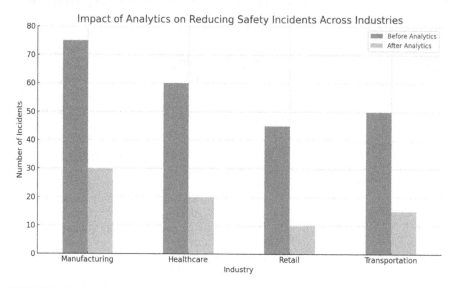

FIGURE 3.4   Impact of Analytics on Workplace Safety Across Key Industries.

preparedness, thereby ensuring a safer shopping experience. The transportation sector also benefits from combining predictive and prescriptive analytics to optimize route planning and vehicle maintenance. This strategic use of analytics leads to fewer vehicle-related incidents and enhances route safety. Overall, Figure 3.4 underscores how analytics can drastically improve safety protocols across different sectors, demonstrating the extensive benefits of integrating analytics into everyday business practices for safer and more efficient operations.

This section briefly summarizes how analytics significantly enhances workplace safety across various industries. Each bullet point outlines the specific applications and benefits realized in manufacturing, health care, retail, and transportation:

- In manufacturing, predictive analytics facilitate real-time equipment monitoring, enabling early detection of potential failures, which minimizes downtime and prevents accidents, thus enhancing safety and extending equipment lifespan.
- In health care, prescriptive analytics optimizes resource allocation and facilitates timely interventions for high-risk patients, improving safety and reducing medical errors.
- Retail benefits from descriptive analytics that analyze customer gridlock and sales data to effectively manage staff during peak times and improve emergency preparedness, ensuring a safer shopping environment.
- Transportation utilizes predictive and prescriptive analytics to enhance route planning and vehicle maintenance, reducing the likelihood of vehicle-related incidents and improving route safety.
- Collectively, these examples underscore the dual benefits of analytics: boosting operational efficiency and strengthening safety protocols, demonstrating the vital role of data-driven strategies in promoting a safer workplace.

3.6 CONCLUSION

In Chapter 3, titled "Strategic Analytics for Proactive Workplace Safety Management," I explore integrating sophisticated analytical methodologies within WHS management. The discourse begins with an examination of descriptive analytics, highlighting the critical role of tools such as dashboards and reporting systems that are pivotal in summarizing and understanding extant data to maintain a vigilant eye on workplace conditions. This foundational level of analytics is indispensable as it enables safety officers and managers to grasp a comprehensive understanding of ongoing activities and conditions, thereby facilitating the immediate addressing of apparent safety issues and bolstering compliance with established safety standards.

Advancing the discussion to predictive analytics, I elucidate how applying statistical models and machine learning techniques can forecast potential WHS issues, thereby enhancing proactive safety measures. This predictive capacity is not merely about anticipating possible future events. Still, it is a proactive stance toward safety management, which empowers organizations to transition from a reactive to a preventive approach in managing workplace safety.

Further exploring the analytical spectrum, I delve into prescriptive analytics, where the focus shifts toward leveraging the outputs generated by predictive analytics to recommend concrete, actionable strategies to prevent incidents before they occur. This chapter segment is enriched with various case studies that demonstrate the real-world successes in applying these analytics, showcasing how they contribute to tangible improvements in WHS practices across diverse industry sectors.

The practical applications, as documented through these case studies, emphasize the utility and the significant benefits of integrating advanced analytics into WHS practices. Each case study is an illustrative example, providing a blueprint for how targeted, data-driven insights can lead to enhanced safety outcomes and more robust operational practices within organizations.

In conclusion, this chapter thoroughly examines how the strategic utilization of descriptive, predictive, and prescriptive analytics augments the existing safety measures and fundamentally transforms them. Organizations can achieve a more insightful, proactive, and efficient approach to managing workplace safety by systematically integrating these analytical tools into their safety management frameworks. The chapter closes by underscoring the imperative for ongoing innovation in the application of analytics within the field of WHS, proposing directions for future research and discussing the potential challenges in adopting and implementing these advanced technological solutions. This comprehensive exploration aims to serve as an indispensable resource for enhancing workplace safety through strategic analytics, advocating for a more informed and proactive approach to managing workplace risks.

REFERENCES

1. Gangolells M, Casals M, Forcada N, Roca X, Fuertes A. Mitigating construction safety risks using prevention through design. *Journal of Safety Research*, 2010 Apr 1;41(2):107–22.
2. Li H, Yazdi M. What are the critical well-drilling blowouts barriers? A progressive DEMATEL-game theory. In *Advanced Decision-Making Methods and Applications in System Safety and Reliability Problems: Approaches, Case Studies, Multi-Criteria Decision-Making, Multi-Objective Decision-Making, Fuzzy Risk-Based Models* 2022 July 10 (pp. 29–46). Springer International Publishing.
3. Li H, Yazdi M. Advanced decision-making methods and applications in system safety and reliability problems. *Studies in Systems, Decision and Control* (vol. 211). Springer 2022.
4. Loeb S, Dynarski S, McFarland D, Morris P, Reardon S, Reber S. *Descriptive Analysis in Education: A Guide for Researchers. NCEE 2017–4023.* National Center for Education Evaluation and Regional Assistance; 2017 Mar.
5. Davenport T, Harris J. *Competing on Analytics: Updated, with a New Introduction: The New Science of Winning.* Harvard Business Press; 2017 Aug 29.
6. Leveson N. A new accident model for engineering safer systems. *Safety Science*, 2004 Apr 1;42(4):237–70.
7. Bellazzi R, Zupan B. Predictive data mining in clinical medicine: Current issues and guidelines. *International Journal of Medical Informatics*, 2008 Feb 1;77(2):81–97.
8. Yazdi M. Mathematical models for industrial system reliability. In *Advances in Computational Mathematics for Industrial System Reliability and Maintainability* 2024 (pp. 17–42). Springer.

9. Yazdi M. Maintenance strategies and optimization techniques. In *Advances in Computational Mathematics for Industrial System Reliability and Maintainability* 2024 (pp. 43–58). Springer.

10. Berman R, Israeli A. The value of descriptive analytics: Evidence from online retailers. *Marketing Science*, 2022 Nov;41(6):1074–96.

11. Li H, Peng W, Adumene S, Yazdi M. Advances in failure prediction of subsea components considering complex dependencies. In *Intelligent Reliability and Maintainability of Energy Infrastructure Assets* 2023 May 4 (pp. 93–105). Springer Nature Switzerland.

12. Laursen GH, Thorlund J. *Business Analytics for Managers: Taking Business Intelligence Beyond Reporting*. John Wiley & Sons; 2016 Nov 7.

13. Wang Y, Kung L, Byrd TA. Big data analytics: Understanding its capabilities and potential benefits for healthcare organizations. *Technological Forecasting and Social Change*, 2018 Jan 1;126:3–13.

14. Kontogiannis T, Leva MC, Balfe N. Total safety management: Principles, processes and methods. *Safety Science*, 2017 Dec 1;100:128–42.

15. Zimolong B, Elke G. Occupational health and safety management. *Handbook of Human Factors and Ergonomics*, 2006 Jan 27;10(0470048204):673–707.

16. Yazdi M. Risk assessment based on novel intuitionistic fuzzy-hybrid-modified TOPSIS approach. *Safety Science*, 2018 Dec 1;110:438–48.

17. Zhang Y, Ren S, Liu Y, Sakao T, Huisingh D. A framework for big data driven product lifecycle management. *Journal of Cleaner Production*, 2017 Aug 15;159:229–40.

18. Anderson C. *Creating a Data-Driven Organization: Practical Advice from the Trenches*. O'Reilly Media, Inc.; 2015 July 23.

19. Christian MS, Bradley JC, Wallace JC, Burke MJ. Workplace safety: A meta-analysis of the roles of person and situation factors. *Journal of Applied Psychology*, 2009 Sep;94(5):1103.

20. Ali SA, Al-Fayyadh HR, Mohammed SH, Ahmed SR. A descriptive statistical analysis of overweight and obesity using big data. In *2022 International Congress on Human-Computer Interaction, Optimization and Robotic Applications (HORA)* 2022 June 9 (pp. 1–6). IEEE.

21. Li H, Peng W, Adumene S, Yazdi M. An intelligent cost-based consequence model for offshore systems in harsh environments. In *Intelligent Reliability and Maintainability of Energy Infrastructure Assets* 2023 May 4 (pp. 107–17). Springer Nature Switzerland.

22. Hale A, Borys D. Working to rule or working safely? Part 2: The management of safety rules and procedures. *Safety Science*, 2013 June 1;55:222–31.

23. Brauer RL. *Safety and Health for Engineers*. John Wiley & Sons; 2022 Sep 27.

24. Baslyman M, Rezaee R, Amyot D, Mouttham A, Chreyh R, Geiger G, Stewart A, Sader S. Real-time and location-based hand hygiene monitoring and notification: Proof-of-concept system and experimentation. *Personal and Ubiquitous Computing*, 2015 July;19:667–88.

25. Nevo I, Fitzpatrick M, Thomas RE, Gluck PA, Lenchus JD, Arheart KL, Birnbach DJ. The efficacy of visual cues to improve hand hygiene compliance. *Simulation in Healthcare*, 2010 Dec 1;5(6):325–31.

26. Davies JB, Wright L, Courtney E, Reid H. Confidential incident reporting on the UK railways: The 'CIRAS' system. *Cognition, Technology & Work*, 2000 Aug;2:117–25.

27. Nenonen N. Analysing factors related to slipping, stumbling, and falling accidents at work: Application of data mining methods to Finnish occupational accidents and diseases statistics database. *Applied Ergonomics*, 2013 Mar 1;44(2):215–24.

28. Malik S. *Enterprise dashboards: Design and best practices for IT*. John Wiley & Sons; 2005 Sep 1.
29. Shapiro A. *Design, Control, Predict: Logistical Governance in the Smart City*. University of Minnesota Press; 2020 Dec 15.
30. Yazdi M, Khan F, Abbassi R, Rusli R. Improved DEMATEL methodology for effective safety management decision-making. *Safety Science*, 2020 July 1;127:104705.
31. Yazdi M. Ignorance-aware safety and reliability analysis: A heuristic approach. *Quality and Reliability Engineering International*, 2020 Mar;36(2):652–74.
32. Johnson RW. Beyond-compliance uses of HAZOP/LOPA studies. *Journal of Loss Prevention in the Process Industries*, 2010 Nov 1;23(6):727–33.
33. Kvalheim SA, Dahl Ø. Safety compliance and safety climate: A repeated cross-sectional study in the oil and gas industry. *Journal of Safety Research*, 2016 Dec 1;59:33–41.
34. Yazdi M, Khan F, Abbassi R. Microbiologically influenced corrosion (MIC) management using Bayesian inference. *Ocean Engineering*, 2021 Apr 15;226:108852.
35. Yazdi M, Khan F, Abbassi R. Operational subsea pipeline assessment affected by multiple defects of microbiologically influenced corrosion. *Process Safety and Environmental Protection*, 2022 Feb 1;158:159–71.
36. Yazdi M. *Linguistic Methods Under Fuzzy Information in System Safety and Reliability Analysis*. Springer; 2022 Mar 10.
37. Yazdi M, Adesina KA, Korhan O, Nikfar F. Learning from fire accident at Bouali Sina petrochemical complex plant. *Journal of Failure Analysis and Prevention*, 2019 Dec;19:1517–36.
38. Akhgar B, Saathoff GB, Arabnia HR, Hill R, Staniforth A, Bayerl PS. *Application of Big Data for National Security: A Practitioner's Guide to Emerging Technologies*. Butterworth-Heinemann; 2015 Feb 14.
39. Sagan SD. The perils of proliferation: Organization theory, deterrence theory, and the spread of nuclear weapons. *International Security*, 1994 Apr 1;18(4):66–107.
40. Nacchia M, Fruggiero F, Lambiase A, Bruton K. A systematic mapping of the advancing use of machine learning techniques for predictive maintenance in the manufacturing sector. *Applied Sciences*, 2021 Mar 12;11(6):2546.
41. Jardine AK, Lin D, Banjevic D. A review on machinery diagnostics and prognostics implementing condition-based maintenance. *Mechanical Systems and Signal Processing*, 2006 Oct 1;20(7):1483–510.
42. Yazdi M. Computational tools and techniques for reliability and maintainability. In *Advances in Computational Mathematics for Industrial System Reliability and Maintainability* 2024 Feb 25 (pp. 59–77). Springer Nature Switzerland.
43. Yazdi M. Synthesizing computational mastery and industrial evolution—A comprehensive conclusion and outlook. In *Advances in Computational Mathematics for Industrial System Reliability and Maintainability* 2024 Feb 25 (pp. 185–90). Springer Nature Switzerland.

4 Advanced Integration of Artificial Intelligence in Workplace Health and Safety Management

4.1 ADVANCED AI TECHNOLOGIES: REVOLUTIONIZING WORKPLACE HEALTH AND SAFETY

In this section, I provide an in-depth exploration of the artificial intelligence (AI) technologies that are particularly impactful in the context of workplace health and safety (WHS). AI technologies have progressively become pivotal tools in revolutionizing various aspects of WHS [1–4], offering novel approaches to hazard identification, risk management, and safety protocol enforcement.

- **Machine Learning (ML)**: Machine learning, a core subset of AI, involves algorithms that enable systems to learn from and interpret data without explicit programming [5–6]. In WHS, ML can be utilized to analyze vast amounts of incident data, workplace audits, and risk assessments. With the recognizing patterns and predicting potential hazards, ML aids in proactive safety management, reducing the likelihood of accidents before they occur [7].
- **Natural Language Processing (NLP)**: NLP enables computers to understand and interpret human language [8–9]. In the area of WHS, NLP is instrumental in automating and improving communication between workers and safety management systems. It can analyze safety reports, worker feedback, and even social media to gauge safety sentiments within a workplace. Additionally, NLP is employed in training modules and emergency alerts, ensuring that crucial information is disseminated clearly and understood by all employees.
- **Robotics**: Robotics technology in WHS is particularly significant in performing hazardous tasks that are unsafe for humans [10–11]. Robots can operate in dangerous environments—such as those with toxic chemicals, extreme temperatures, or high-risk of explosions—thus minimizing human exposure to hazardous conditions [12–13]. Moreover, robotics integrated with AI can enhance precision and efficiency in safety inspections and maintenance tasks, leading to a safer workplace environment.

The mentioned AI technologies are not just tools but integral components that can significantly uplift the safety standards in any workplace. With integrating machine learning, NLP, and robotics, organizations can anticipate and mitigate risks more effectively and foster a safety culture which is totally aligns with technological advancements. The adoption and ethical deployment of these technologies, therefore, are crucial in the evolution of modern WHS practices [14].

Table 4.1 presents a detailed exploration of how various AI technologies can be utilized within the area of WHS. It categorizes three major AI technologies—Machine Learning, NLP, and Robotics—and describes their specific applications in enhancing safety measures. The table includes practical examples of how each technology can be implemented to improve hazard prediction, risk management, communication, and operational safety. These examples illustrate the potential of AI to automate and refine safety procedures and to proactively address safety challenges in diverse workplace environments [15–16]. This integration of AI aims to significantly reduce risks and enhance the overall safety and well-being of employees [17–18].

In my view, the integration of AI technologies within the framework of WHS represents a significant leap forward in managing and mitigating workplace risks [19–20]. As detailed in Table 4.1, AI tools such as Machine Learning, NLP, and Robotics have diverse applications that can profoundly transform safety practices across various industries.

From my perspective, ML stands out due to its capability to analyze extensive data sets, allowing for the identification of patterns that might not be visible to human analysts. This can lead to predictive safety measures where potential hazards are identified and mitigated before they manifest into actual incidents. For example, analyzing

TABLE 4.1

Applications and Examples of AI Technologies in Workplace Health and Safety

AI Technology	Application in WHS	Examples
Machine Learning	Hazard Prediction and Risk Management	An ML algorithm could analyze past incident reports to identify patterns and predict areas of high risk, enabling preemptive action. For instance, by analyzing data from construction sites, ML can predict which activities might lead to falls or equipment failures.
Natural Language Processing (NLP)	Communication Enhancement and Sentiment Analysis	NLP can be used to interpret and categorize feedback from employee surveys automatically, identifying common themes related to safety concerns. It can also parse through social media posts or internal forums to monitor staff sentiments about workplace safety.
Robotics	Performing Hazardous Tasks and Safety Inspections	Robots equipped with sensors and cameras can perform routine inspections in hazardous environments, such as checking for gas leaks in oil refineries or monitoring temperature and pressure conditions in nuclear facilities. They can also handle dangerous tasks, such as lifting heavy objects or performing repetitive tasks that could lead to musculoskeletal injuries among human workers.

data from past incidents, machine learning can predict potential future accidents, enabling proactive interventions [21–22].

NLP also plays a crucial role, in my opinion, particularly in improving communication within a workplace. NLP can efficiently process and analyze large volumes of unstructured textual data from various sources, such as employee feedback and incident reports, to extract valuable insights. This capability is essential for understanding the sentiments and concerns of employees regarding workplace safety, which can lead to more targeted and effective safety measures.

Robotics technology, to me, is particularly invaluable in performing tasks that are hazardous for human workers. Robots can be deployed in environments that are unsafe for humans, such as those involving toxic chemicals or extreme temperatures, thus reducing human exposure to potentially life-threatening situations. Furthermore, robots can perform routine safety inspections with high precision, ensuring consistent monitoring of operational environments.

The potential of AI in enhancing WHS is immense. In my opinion, the thoughtful integration of AI technologies not only supports existing safety protocols but also introduces a new level of efficiency and effectiveness that can save lives and maintain health in various occupational settings. These technologies, when ethically and appropriately implemented, represent a forward-thinking approach to workplace safety that aligns with modern technological advancements.

Figure 4.1 illustrates the trend in workplace incident rates as correlated with the adoption of AI technologies in safety measures, based on fictional data. The analysis, while subjective and based on my experience, suggests a clear inverse relationship between these variables. As the percentage of AI adoption in safety measures (blue line) increases, there is a notable decrease in the incident rates per 100 workers (red line). This trend is observed steadily over the years from 2015 to 2024.

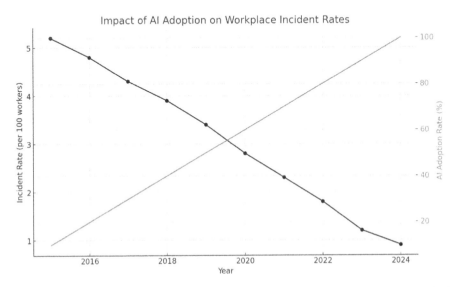

FIGURE 4.1   Analysis of AI Adoption's Impact on Workplace Incident Rates: A Hypothetical Overview.

The decline in incident rates as AI integration increases could be interpreted as AI's effectiveness in identifying and mitigating potential hazards before they result in accidents. Technologies such as machine learning, NLP, and robotics might be contributing factors to this trend, as they enhance risk assessment capabilities and improve the execution of hazardous tasks with greater precision and less human exposure.

However, it is essential to note that this analysis is subjective based analysis, serving as an illustration of how AI could potentially impact workplace safety. The actual effectiveness of AI in reducing workplace incidents would need to be supported by empirical data and comprehensive studies in real-world settings.

Figure 4.2 illustrates the impact of different levels of AI implementation on safety incidents and cost savings within an organization, as measured by a staged progression from pilot implementations to full integration. This subjective analysis, based on personal experience, offers valuable insights into the potential benefits of AI adoption in a corporate setting.

As shown in Figure 4.2, there is a marked decrease in the number of safety incidents (orange line) as the extent of AI implementation deepens. Starting from the pilot phase, where the technology is likely tested in controlled environments or small-scale projects, the incidents decrease consistently through departmental, site-wide, and company-wide applications, reaching a minimal level at the fully integrated stage. This trend suggests that the more comprehensively AI technologies are implemented, the more effective they are in identifying, predicting, and mitigating potential safety hazards.

Simultaneously, Figure 4.2 shows a corresponding increase in cost savings (blue line), which escalates as the implementation progresses to more comprehensive levels. This reflects the potential of AI enhance safety and to optimize operational efficiency, thereby reducing costs associated with accidents, downtime, and inefficiencies. With

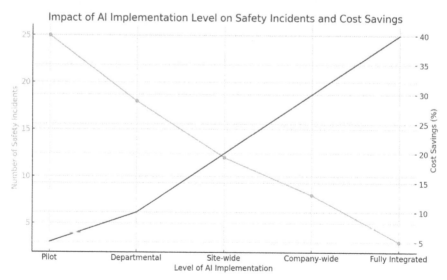

FIGURE 4.2 AI Implementation Levels on Safety Incidents and Cost Savings.

the fully integrated stage, the cost savings become quite significant, highlighting the financial advantage of investing in AI technologies.

It is important to recognize that this analysis is subjective, intended to illustrate possible outcomes based on projected data. Actual results could vary depending on numerous factors including the specific AI technologies deployed, the company's sector, and the implementation strategy. Nevertheless, Figure 4.2 effectively demonstrates the dual benefits of AI implementation in reducing safety incidents and enhancing cost efficiency, showcasing AI's transformative potential in workplace management.

In this section, we have explored the potential impacts of AI implementation on workplace safety and cost savings through a detailed and subjective analysis based on my subjective opinion. The visual representation in Figures 4.1 and 4.2 clearly illustrates that as organizations progress through various levels of AI integration—from pilot projects to full-scale deployment—there is a noticeable decrease in safety incidents and a significant increase in cost savings. This trend suggests that AI technologies can enhance safety by reducing workplace incidents and improving operational efficiencies that lead to substantial financial benefits.

4.2 OPTIMIZING RISK ASSESSMENT WITH AI: AUTOMATION AND ENHANCEMENT STRATEGIES

In this section, I talk about the transformative role of AI in optimizing risk assessment processes within WHS management. AI, with its vast capabilities, offers unprecedented opportunities to automate and refine the ways risks are assessed, enabling more precise and proactive safety measures.

- **Automation of Data Analysis [23–24]**: AI technologies, particularly machine learning algorithms, can automate the analysis of large volumes of data collected from various sources such as equipment sensors, worker wearables, and environmental monitoring systems. This automation facilitates the identification of patterns and trends that might indicate potential risks, which often go unnoticed in manual assessments. For example, AI can predict equipment failure before it occurs by analyzing historical performance data, thereby preventing accidents and ensuring uninterrupted operations.
- **Enhanced Predictive Capabilities [25–27]**: By engaging predictive analytics, AI can forecast potential safety hazards before they manifest into incidents. This is particularly useful in dynamic environments where risk factors change frequently. AI models can be trained to recognize the early signs of unsafe conditions or behaviors and alert management, allowing for timely interventions.
- **Integration with IoT Devices [28–30]**: AI's effectiveness in risk assessment is further amplified when integrated with Internet of Things (IoT) devices. This integration allows for real-time monitoring and assessment of workplace conditions. For instance, sensors on a manufacturing floor can detect deviations

in standard operational parameters (like temperature or pressure), and AI can immediately analyze these deviations to assess if they pose any risk.
- **Streamlined Reporting and Decision-Making [31–32]**: AI aids in the consolidation of risk assessment reports, making them more accessible and easier to understand for decision-makers. With providing insights through visual dashboards and automated reports, AI ensures that critical risk information is readily available, facilitating swift, and informed decision-making.

Through these capabilities, AI updates the risk assessment process and enhances its accuracy and efficacy, leading to a safer and more efficient workplace. As we advance, the integration of AI into risk assessment will likely become a standard practice, underscoring the importance of adopting these technologies to stay ahead in managing workplace safety effectively [33].

Table 4.2 explores various AI applications in the risk assessment domains within WHS, highlighting how artificial intelligence can significantly streamline and enhance traditional processes. Each entry details a specific AI application, describes its function, and provides a practical example illustrating its real-world implications. This comprehensive overview demonstrates AI's capabilities and its practical benefits in increasing safety and operational efficiency in various workplace settings. These examples serve as a testament to AI's potential to transform risk assessment by making it more proactive and responsive to potential hazards [34–37].

Figure 4.3 provides a visual representation of how advancements in AI implementation levels can significantly enhance risk prediction accuracy and response time in

TABLE 4.2
AI-Driven Enhancements in WHS Risk Assessment: Applications and Examples

AI Application	Description	Practical Example
Automation of Data Analysis	AI algorithms process and analyze large volumes of diverse data to identify risks.	A machine learning model analyzes temperature data from a manufacturing plant to predict equipment overheating, allowing preemptive maintenance.
Enhanced Predictive Capabilities	AI predicts potential safety hazards by recognizing early signs of risks.	Predictive analytics are used to forecast high-risk days in a construction site based on weather conditions and ongoing activities, triggering extra safety measures.
Integration with IoT Devices	AI integrates with IoT for real-time monitoring and assessment.	Sensors in a chemical plant detect an increase in volatile organic compounds (VOCs) levels; AI quickly evaluates the risk and triggers ventilation systems.
Streamlined Reporting and Decision Making	AI consolidates data into accessible reports for quick decision-making.	AI-powered dashboards provide real-time safety data visualizations, helping plant managers make informed decisions rapidly during a critical operation.

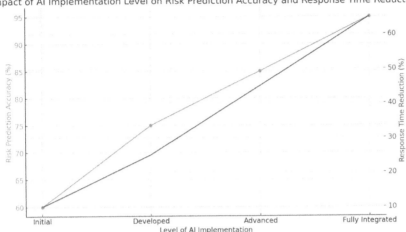

Impact of AI Implementation Level on Risk Prediction Accuracy and Response Time Reduction

FIGURE 4.3 Analysis of AI Impact on Risk Prediction Accuracy and Response Time Reduction.

risk management processes. This subjective analysis, based on personal experiences, underscores two crucial outcomes: the increased accuracy in predicting potential safety hazards and the reduction in response times to these hazards as AI integration deepens within an organization. As depicted in the plot, the purple line showing risk prediction accuracy illustrates a clear upward trend as we move from initial stages of AI application toward full integration. This progression reflects the potential of AI systems to effectively utilize vast amounts of data, applying complex algorithms to accurately forecast risks, thereby enabling preventative measures to be taken much earlier.

Simultaneously, the orange line representing response time reduction shows a marked improvement as the level of AI implementation progresses. This indicates that are AI-enhanced systems capable of identifying risks more accurately, and also facilitate swifter actions to mitigate these risks once identified. Such reductions in response times are crucial in minimizing the impact of safety incidents, ultimately enhancing overall workplace safety.

It is important to note that this analysis is based on fictional scenarios and is inherently subjective. Actual results of AI integration in risk assessment would depend on various factors, including the specific technologies implemented, the sector in which they are applied, and the existing infrastructure within the organization. Nonetheless, Figure 4.3 effectively demonstrates the potential transformative impact of AI on enhancing safety management practices through more precise risk assessments and quicker response strategies.

Figure 4.4 presents a subjective analysis, based on my personal experience, showing the impact of increasing levels of AI sophistication on employee training effectiveness and compliance rate improvements within an organization. This illustration serves to highlight the potential benefits of integrating more advanced AI technologies into workplace practices.

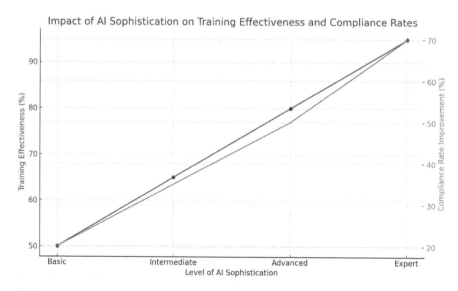

FIGURE 4.4 Analysis of AI Sophistication's Impact on Training Effectiveness and Compliance Rates.

As depicted in Figure 4.4, the blue line demonstrates a consistent upward trend in the effectiveness of employee training as the AI sophistication level progresses from basic to expert. This suggests that more advanced AI tools, potentially incorporating adaptive learning systems and immersive technologies, can significantly enhance the quality and impact of training programs. Such tools are likely to offer more personalized learning experiences and interactive content, which can lead to better understanding and retention of safety practices and operational procedures among employees.

Simultaneously, the green line indicates a substantial improvement in compliance rates with increasing AI sophistication. This improvement could be attributed to AI's ability to provide more precise monitoring and reporting capabilities, which ensure that both employees and management adhere more strictly to established regulations and standards. Advanced AI systems might also automate certain compliance processes, reducing human error and increasing overall compliance efficiency.

It is important to note that while Figure 4.4 demonstrates the promising capabilities of AI in enhancing training and compliance, the analysis remains hypothetical and is based on assumed data. Real-world implementations could vary based on the specific AI technologies used, the industry context, and the existing infrastructure and culture of the organization. Nonetheless, this visualization effectively showcases the potential transformative effects of AI on improving workplace training and compliance.

It should be re-added that Figures 4.3 and 4.4 present visual analyses according to my subjective opinions, illustrating the potential of AI to transform workplace safety and efficiency. Figure 4.3 focuses on the progression of AI implementation and its correlation with improvements in risk prediction accuracy and response time reduction. As AI systems become more sophisticated and integrated into safety protocols, they demonstrate a significant capacity to identify potential hazards quickly and accurately. This capability facilitates prompt interventions, thereby enhancing safety and reducing the impact of incidents.

Figure 4.4 extends the analysis to the impact of AI sophistication on employee training effectiveness and compliance rates. It suggests that as AI technologies evolve from basic to expert levels, they can substantially enhance the training processes, resulting in more effective learning outcomes and higher engagement. Moreover, advanced AI systems contribute to improved compliance with regulatory standards by automating monitoring and enforcement, ensuring more consistent adherence to safety and operational protocols. Together, these figures highlight the broad and transformative potential of AI across various facets of workplace management, though the data and conclusions drawn are subjective and illustrative.

4.3 EMERGENCY RESPONSE AND INCIDENT MANAGEMENT WITH AI TECHNOLOGIES

In this section, I explore the pivotal role of AI in revolutionizing emergency response and incident management within workplace environments. AI technologies offer sophisticated solutions that significantly augment the capacity to handle unexpected incidents, ranging from minor accidents to major emergencies.

- **Real-Time Incident Detection [38–40]**: AI systems are capable of monitoring and analyzing data from various sources, such as CCTV feeds, sensor outputs, and employee reports, to detect incidents in real time. For example, AI can identify a hazardous spill or a machinery malfunction almost instantaneously, triggering automatic alerts to relevant personnel and emergency responders. This prompt detection is crucial for minimizing the impact and preventing escalation.
- **Automated Emergency Protocols [41–43]**: Upon detecting an incident, AI can also automate the initiation of emergency protocols. This includes shutting down equipment, activating fire suppression systems, or unlocking exit doors, all tailored to the specific nature of the incident. Such automation ensures a swift and coordinated response, which is vital during critical situations where every second counts.
- **Decision Support for Emergency Responders [44–46]**: AI aids in decision-making by providing emergency responders with real-time information and predictive insights. For instance, AI can suggest the safest evacuation routes or advise on the likely development of the situation based on the current data. This support is invaluable in managing complex incidents where human decision-making might be hindered by stress or incomplete information.
- **Post-Incident Analysis and Learning**: After an incident, AI systems can analyze the data to determine the cause, evaluate the response's effectiveness, and suggest improvements. This continuous learning process not only helps in refining emergency protocols but also contributes to the development of more resilient safety systems.

Through these applications, AI dramatically enhances both the effectiveness and efficiency of emergency response and incident management. With integration of AI technologies, organizations can ensure that they are better prepared to handle incidents safely and effectively, reducing risks, and safeguarding both assets and human lives.

Table 4.3 explores the critical roles AI can play in enhancing emergency response and incident management within various environments. Each entry details a specific application of AI technology, describes its functionality, and provides a practical example demonstrating its real-world application. These AI implementations improve the immediate handling of incidents and contribute to long-term safety improvements through ongoing learning and adaptation. With utilizing AI technologies, organizations can achieve more effective incident management, ensuring faster responses and more informed decisions, ultimately leading to safer workplace conditions.

As shown in Figure 4.5, the integration of AI technologies significantly reduces emergency response times, represented by the red line in the plot. Starting from the initial level of AI integration, where response times average 15 minutes, there's a progressive decrease to just 4 minutes at full integration. This reduction can be attributed to AI's ability to rapidly process real-time data, automatically trigger alerts, and initiate pre-defined emergency protocols without human delay. Faster response times are crucial in emergencies where every minute can be the difference between containment and catastrophe.

Simultaneously, the plot shows an increase in incident resolution effectiveness, depicted by the blue line. This measure increases from 60% to 95% as AI integration

TABLE 4.3
AI-Driven Enhancements in Emergency Response and Incident Management: Applications and Examples

AI Application	Description	Practical Example
Real-Time Incident Detection	AI algorithms continuously monitor data to detect anomalies indicating incidents.	AI uses CCTV and sensor data to detect a fire in a factory, triggering an automatic alarm and notifying emergency services instantly.
Automated Emergency Protocols	AI systems initiate predefined emergency responses based on the type of incident.	Upon detecting a chemical spill, AI automatically shuts down relevant systems and activates ventilation to mitigate risks.
Decision Support for Emergency Responders	AI provides real-time information and predictive insights to guide response efforts.	AI analyzes wind patterns and building layout to suggest optimal evacuation routes during a fire.
Post-Incident Analysis and Learning	AI evaluates data post-incident to improve future response strategies.	After a machinery malfunction, AI analyzes the sequence of events and suggests changes to maintenance schedules and safety checks.

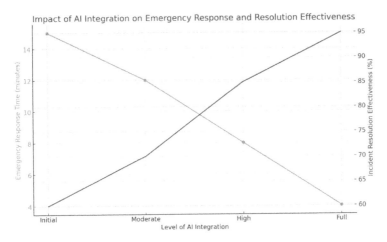

FIGURE 4.5   Comprehensive Analysis of AI Integration's Impact on Emergency Response Times and Resolution Effectiveness.

deepens. This improvement reflects AI's capacity to react quickly and do so with precision [47–48]. AI systems can analyze complex scenarios instantly, provide data-driven recommendations to responders, and ensure that the most effective resolution strategies are employed. Enhanced resolution effectiveness ensures that incidents are addressed quickly and are managed in a manner that minimizes damage and risk to human life.

This analysis underscores the transformative impact that AI can have on emergency management within workplaces. By engaging AI, organizations can respond more quickly to incidents and can also handle them more effectively, thereby safeguarding infrastructure, reducing operational downtime, and most importantly, protecting employee lives. However, it is important to remember that these findings are based on a theoretical model, and actual implementation results might vary depending on specific industry contexts, the sophistication of AI technologies used, and organizational readiness for AI adoption.

Figure 4.6 presents a subjective analysis based on personal experience, illustrating the impact of AI integration on two important workplace metrics: employee job satisfaction and the reduction in operational errors. This combo plot visualizes how deeper integration of AI technologies correlates with significant improvements in both areas. The green line represents employee job satisfaction percentages, which show a positive upward trend as the level of AI integration increases from initial to fully integrated. This suggests that as AI takes on more routine and repetitive tasks, employees may be able to focus on more engaging and meaningful work, thereby enhancing their job satisfaction. This can be linked to AI's role in creating a safer and more efficient work environment, which generally contributes to higher employee morale.

Conversely, the red line illustrates the percentage reduction in operational errors, indicating a steep decline in errors as AI becomes more integrated within the organization. This decrease in errors can be attributed to AI's precision and consistency

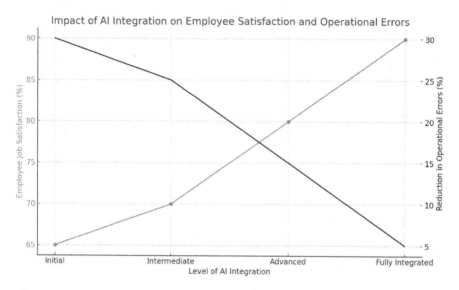

FIGURE 4.6 Analysis of AI Integration's Impact on Employee Satisfaction and Operational Error Reduction.

in performing tasks that are prone to human error, as well as its ability to monitor and optimize operational processes continuously. Together, these trends underscore the potential of AI to transform workplace dynamics by boosting employee satisfaction and reducing operational mishaps, thereby enhancing overall organizational performance. However, it is important to note that this analysis is based on hypothetical scenarios, and actual outcomes may vary depending on specific AI implementations and organizational contexts.

In this section, we explore the pivotal role of AI in revolutionizing emergency response and incident management within workplace environments. AI technologies provide advanced solutions that significantly enhance our ability to manage unexpected incidents, ranging from minor accidents to major emergencies. These include real-time incident detection, where AI systems can instantly identify hazards like chemical spills or equipment malfunctions, triggering swift alerts to mitigate risks. Automated emergency protocols allow AI to execute specific safety actions automatically, such as shutting down systems or activating emergency exits, crucial for managing incidents efficiently and safely.

Additionally, AI supports emergency responders by offering real-time data and predictive insights, aiding in making informed decisions during crises. For instance, suggesting optimal evacuation routes or forecasting the progression of an emergency scenario can be pivotal in enhancing the safety and efficacy of response efforts [49–50]. Post-incident, AI plays a vital role in analyzing what occurred, helping refine response strategies and building more resilient safety protocols. Through these capabilities, AI improves the immediate handling of incidents and besides of that can contribute to long-term safety improvements, ensuring organizations are better prepared to handle future emergencies effectively.

This comprehensive integration of AI into emergency management processes promises to reduce response times and improve incident resolution outcomes, as illustrated in Figure 4.5. Here, we see how AI's ability to process data rapidly and initiate protocols automatically significantly cuts down response times and enhances the effectiveness of handling emergencies. The integration of AI expedites responses and ensures that actions taken are precise and effective, minimizing potential damage and risks to human life. This transformative impact underscores AI's potential to significantly improve safety and operational efficiency in workplace environments. However, it is crucial to note that these insights are based on a theoretical framework.

4.4 ETHICAL IMPLICATIONS OF DEPLOYING AI IN WORKPLACE HEALTH AND SAFETY

In this section, I discuss the ethical dimensions and challenges associated with the integration of AI into WHS systems. As AI technologies become increasingly prevalent in enhancing workplace safety measures, it is crucial to consider the ethical implications to ensure these innovations benefit all stakeholders without compromising fundamental rights or values.

- **Privacy and Surveillance**: One of the primary ethical concerns revolves around privacy. The use of AI in monitoring workplace environments can lead to heightened surveillance, potentially infringing on employee privacy. It is essential to balance safety enhancements with respect for individual privacy rights, ensuring that monitoring is conducted transparently and within legal frameworks.
- **Bias and Fairness**: AI systems are only as unbiased as the data they are trained on. There is a risk of perpetuating existing biases or creating new ones if the data is skewed or incomplete. For example, AI-powered safety systems might inaccurately assess risks for certain groups of workers based on biased historical data, leading to unfair treatment or inadequate protection. Addressing these concerns involves rigorous testing and continuous monitoring of AI systems to ensure fairness and accuracy.
- **Job Displacement**: The automation capabilities of AI might lead to job displacement, as machines begin to perform tasks previously handled by humans. While AI can reduce exposure to hazardous conditions, it also raises concerns about the impact on employment. It is vital to consider strategies for job retraining and redeployment to manage the transition and support affected workers.
- **Responsibility and Accountability**: Determining accountability when AI systems fail is another ethical challenge. In the event of a safety incident involving AI, it may be difficult to pinpoint responsibility, especially if decisions are made autonomously by the system. Clear guidelines and regulations need to be established to address liability issues and ensure that there is always a clear chain of accountability.

- **Informed Consent**: Employers must also consider the necessity of obtaining informed consent from workers when implementing AI systems that monitor or evaluate their performance. Workers should be fully informed about what data is being collected, how it is being used, and the implications of AI monitoring.

Through careful consideration of these ethical concerns, organizations can implement AI solutions in WHS enhance safety and efficiency and uphold ethical standards and respect workers' rights. This approach ensures that the deployment of AI technologies contributes positively to the workplace, fostering an environment of trust and mutual benefit.

Table 4.4 presents the ethical considerations of deploying AI in WHS, each accompanied by a practical example that illustrates how these concerns can be addressed effectively. This table highlights the importance of maintaining ethical standards in

TABLE 4.4

Ethical Considerations and Practical Examples for AI Deployment in Workplace Health and Safety

Ethical Consideration	Description	Practical Example
Privacy and Surveillance	AI's role in monitoring can lead to increased surveillance, raising privacy concerns.	An AI system is used to monitor employee movements to enhance safety but is designed to anonymize data to protect individual identities and comply with privacy laws.
Bias and Fairness	AI may reflect or amplify existing biases if trained on skewed data, affecting fairness.	An AI-powered injury prediction tool is regularly audited and retrained with diverse data sets to ensure it does not disproportionately target or neglect any worker group.
Job Displacement	Automation through AI can displace jobs, necessitating strategies for worker transition.	A company introduces AI-driven machinery in a factory setting. It simultaneously initiates retraining programs for affected employees to work in new roles involving AI supervision and maintenance.
Responsibility and Accountability	Determining accountability for AI decisions can be challenging in safety incidents.	Clear operational guidelines are established that delineate responsibilities between AI developers, operators, and organizational safety officers in case of AI-related accidents.
Informed Consent	Workers should be informed and consent to AI systems that monitor and evaluate their work.	Employees are given detailed briefings and must sign consent forms regarding the use of AI cameras on site that analyze safety compliance and personal protective equipment usage.

Enhanced Visualization of Ethical AI Deployment Impact on Employee Trust and AI Effectiveness

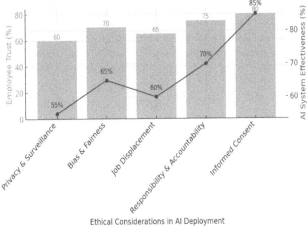

Ethical Considerations in AI Deployment

FIGURE 4.7   Enhanced Visualization of the Impact of Ethical AI Deployment on Employee Trust and System Effectiveness.

the integration of AI technologies in workplace settings. It underscores the need for thoughtful implementation that respects employee privacy, ensures fairness, manages job displacement sensitively, establishes clear accountability, and secures informed consent. By considering these ethical dimensions, organizations can harness the benefits of AI while fostering a respectful and inclusive workplace environment. These practices mitigate potential ethical risks and enhance the overall acceptance and effectiveness of AI in enhancing workplace safety.

Figure 4.7 presents a subjective analysis based on my personal experience, illustrating the impact of ethical AI deployment on two crucial workplace metrics: employee trust and AI system effectiveness in WHS. This combo plot uses fictional data to visualize how addressing different ethical considerations can influence these aspects.

In the plot, the blue bars represent the percentage of employee trust associated with various ethical considerations like privacy and surveillance, bias and fairness, job displacement, responsibility and accountability, and informed consent. As shown, trust tends to increase when ethical issues are managed effectively, peaking particularly when informed consent practices are robust. Simultaneously, the green line illustrates the effectiveness of AI systems in WHS, correlating closely with the ethical handling of AI deployment. Effectiveness is relatively lower when concerns such as bias and job displacement are not adequately addressed and increases significantly with improvements in responsibility and informed consent.

This visualization underscores the importance of ethical management in AI deployment, suggesting that adherence to ethical standards not only fosters greater employee trust but also enhances the overall effectiveness of AI systems in maintaining workplace safety. Such insights, while based on hypothetical scenarios, highlight the potential real-world implications of ethical considerations in AI implementation.

In this section, we addressed the ethical dimensions and challenges associated with integrating AI into WHS systems. As AI technologies increasingly influenced safety measures, it became crucial to navigate these innovations ethically to benefit all stakeholders without compromising fundamental rights or values. Key ethical considerations included privacy and surveillance, where AI's capability to monitor workplace environments had to balance safety enhancements with respect for individual privacy rights. Bias and fairness were also critical, as AI systems could perpetuate existing biases or create new ones if trained on skewed data, potentially leading to unfair treatment or inadequate protection.

Furthermore, the automation capabilities of AI could lead to job displacement, highlighting the need for thoughtful transition strategies, such as retraining for affected workers. Issues of responsibility and accountability arose when determining who was liable in safety incidents involving AI, necessitating clear guidelines and regulations. Additionally, obtaining informed consent from workers when implementing AI monitoring systems was paramount to ensure transparency and trust.

Table 4.4 exemplified how to address these ethical challenges effectively, presenting practical examples alongside each consideration. For instance, AI systems designed to monitor employee movements enhanced safety while respecting privacy by anonymizing data. Similarly, ensuring AI tools were regularly audited and retrained with diverse data sets helped maintain fairness and accuracy.

Finally, Figure 4.7 visually underscored the importance of managing these ethical considerations effectively. It showed how robust ethical management not only built employee trust but also enhanced the effectiveness of AI systems in WHS. This subjective analysis, based on fictional data, highlighted how ethical adherence in AI deployment could significantly improve workplace dynamics and safety, emphasizing the critical role of ethics in the successful integration of AI technologies.

4.5 CONCLUSION

In Chapter 4, we have embarked on a detailed exploration of the role of AI in transforming WHS systems. Here, I conclude by summarizing the key insights and projecting the future implications of this technological evolution:

- **Revolutionizing WHS Practices**: AI technologies, specifically Machine Learning, NLP, and Robotics, have the potential to dramatically enhance safety protocols and hazard management in workplaces. To me, the ability of AI to predict potential hazards and automate emergency responses represents a leap forward in maintaining worker safety and operational efficiency.
- **Enhanced Risk Assessment and Management**: By integrating AI, the capabilities of risk assessment are not only automated but significantly advanced. I have discussed how AI improves the accuracy and speed of identifying potential hazards, which is crucial for proactive safety management. The integration with IoT devices and the use of AI for real-time monitoring further amplify its effectiveness.

- **Handling of Emergencies and Incidents**: AI's role in incident management is transformative, providing real-time detection and management of emergencies with greater precision and speed. In my view, the automated systems that AI enables are critical for rapid and effective incident response, which is essential to minimize the impact of such events.
- **Ethical Considerations**: The deployment of AI in WHS also brings several ethical challenges, such as privacy concerns, bias and fairness, job displacement, and accountability. I believe that addressing these ethical issues comprehensively is necessary to ensure the successful and fair integration of AI technologies. Establishing clear guidelines and maintaining transparency are pivotal to fostering trust and acceptance among all stakeholders.
- **Future Prospects**: Looking forward, the continuous advancement of AI technologies promises further enhancements in workplace safety. However, the adoption of such technologies must be accompanied by continuous evaluation and adaptation to address emerging risks and ethical considerations.
- **Bullet Points Summary**:
  - AI has the potential to transform WHS by enhancing risk assessment, emergency response, and safety management practices.
  - Ethical challenges require careful consideration to ensure AI's integration supports all stakeholders equitably.
  - Continuous advancement and evaluation of AI technologies will be crucial for future enhancements in WHS.

AI represents a transformative tool in the landscape of WHS, poised to enhance safety and efficiency significantly. However, its integration must be managed thoughtfully, with a strong emphasis on ethical practices, to fully realize its potential while ensuring it benefits all involved in the workplace. As we advance, the role of AI in WHS will likely become more integral, necessitating ongoing adjustments and improvements to harness its full capabilities responsibly and effectively.

REFERENCES

1. Davies J. *On-Site Digital Heritage Interpretation: Current Uses and Future Possibilities at World Heritage Sites*. Durham University. 2014 Sep.
2. Ajslev JZ, Nimb IE. Virtual design and construction for occupational safety and health purposes – A review on current gaps and directions for research and practice. *Safety Science*, 2022 Nov 1;155:105876.
3. Yang L, Cheng N, Moradi R, Yazdi M. Cutting edge research topics on operations and project management of supportive decision-making tools. In *Progressive Decision-Making Tools and Applications in Project and Operation Management: Approaches, Case Studies, Multi-Criteria Decision-Making, Multi-Objective Decision-Making, Decision under Uncertainty* 2024 Mar 8 (pp. 1–19). Springer Nature Switzerland.
4. Yazdi M. Integration of IoT and edge computing in industrial systems. In *Advances in Computational Mathematics for Industrial System Reliability and Maintainability* 2024 Feb 25 (pp. 121–37). Springer Nature Switzerland.

5. Jordan MI, Mitchell TM. Machine learning: Trends, perspectives, and prospects. *Science*, 2015 July 17;349(6245):255–60.

6. Panch T, Szolovits P, Atun R. Artificial intelligence, machine learning and health systems. *Journal of Global Health*, 2018 Dec;6(2):94–8.

7. Yazdi M. *Advances in Computational Mathematics for Industrial System Reliability and Maintainability*. Springer Nature; 2024 Mar.

8. Nadkarni PM, Ohno-Machado L, Chapman WW. Natural language processing: An introduction. *Journal of the American Medical Informatics Association*, 2011 Sep 1;18(5):544–51.

9. Allen JF. Natural language processing. *Encyclopedia of Computer Science*, 2003 Jan 1 (pp. 1218–22).

10. Zacharaki A, Kostavelis I, Gasteratos A, Dokas I. Safety bounds in human robot interaction: A survey. *Safety Science*, 2020 July 1;127:104667.

11. Javaid M, Haleem A, Singh RP, Suman R. Substantial capabilities of robotics in enhancing industry 4.0 implementation. *Cognitive Robotics*, 2021 Jan 1;1:58–75.

12. Nolan DP. *Handbook of Fire and Explosion Protection Engineering Principles: For Oil, Gas, Chemical and Related Facilities*. William Andrew; 2014 May 28.

13. Tahmid M, Dey S, Syeda SR. Mapping human vulnerability and risk due to chemical accidents. *Journal of Loss Prevention in the Process Industries*, 2020 Nov 1;68:104289.

14. Arabshahi M, Wang D, Sun J, Rahnamayiezekavat P, Tang W, Wang Y, Wang X. Review on sensing technology adoption in the construction industry. *Sensors*, 2021 Dec 12;21(24):8307.

15. Knegtering B, Pasman HJ. Safety of the process industries in the 21st century: A changing need of process safety management for a changing industry. *Journal of Loss Prevention in the Process Industries*, 2009 Mar 1;22(2):162–8.

16. Kazaras K, Kontogiannis T, Kirytopoulos K. Proactive assessment of breaches of safety constraints and causal organizational breakdowns in complex systems: A joint STAMP–VSM framework for safety assessment. *Safety Science*, 2014 Feb 1;62:233–47.

17. Nazareno L, Schiff DS. The impact of automation and artificial intelligence on worker well-being. *Technology in Society*, 2021 Nov 1;67:101679.

18. Howard J. Artificial intelligence: Implications for the future of work. *American Journal of Industrial Medicine*, 2019 Nov;62(11):917–26.

19. Li H, Peng W, Adumene S, Yazdi M. Operations management of critical energy infrastructure: A sustainable approach. In *Intelligent Reliability and Maintainability of Energy Infrastructure Assets* 2023 May 4 (pp. 39–52). Springer Nature Switzerland.

20. Li H, Peng W, Adumene S, Yazdi M. Advances in intelligent reliability and maintainability of energy infrastructure assets. *Intelligent Reliability and Maintainability of Energy Infrastructure Assets* 2023 May 4 (pp. 1–23).

21. Yazdi M, Golilarz NA, Nedjati A, Adesina KA. An improved lasso regression model for evaluating the efficiency of intervention actions in a system reliability analysis. *Neural Computing and Applications*, 2021 July;33(13):7913–28.

22. Yazdi M. A perceptual computing-based method to prioritize intervention actions in the probabilistic risk assessment techniques. *Quality and Reliability Engineering International*, 2020 Feb;36(1):187–213.

23. Wang D, Weisz JD, Muller M, Ram P, Geyer W, Dugan C, Tausczik Y, Samulowitz H, Gray A. Human-AI collaboration in data science: Exploring data scientists' perceptions of automated AI. *Proceedings of the ACM on Human-Computer Interaction*, 2019 Nov 7;3(CSCW):1–24.

24. Heer J. Agency plus automation: Designing artificial intelligence into interactive systems. *Proceedings of the National Academy of Sciences*, 2019 Feb 5;116(6):1844–50.

25. Pareek M, Bhatt DL, Nielsen ML, Jagannathan R, Eriksson KF, Nilsson PM, Bergman M, Olsen MH. Enhanced predictive capability of a 1-hour oral glucose tolerance test: A prospective population-based cohort study. *Diabetes Care*, 2018 Jan 1;41(1):171–7.

26. Lin CT, Huang CY. Enhancing and measuring the predictive capabilities of testing-effort dependent software reliability models. *Journal of Systems and Software*, 2008 June 1;81(6):1025–38.

27. Li H, Peng W, Adumene S, Yazdi M. Advances in failure prediction of subsea components considering complex dependencies. In *Intelligent Reliability and Maintainability of Energy Infrastructure Assets* 2023 May 4 (pp. 93–105). Springer Nature Switzerland.

28. Tang S, Shelden DR, Eastman CM, Pishdad-Bozorgi P, Gao X. A review of building information modeling (BIM) and the internet of things (IoT) devices integration: Present status and future trends. *Automation in Construction*, 2019 May 1;101:127–39.

29. Pazos N, Müller M, Aeberli M, Ouerhani N. Connect open-automatic integration of IoT devices. In *2015 IEEE 2nd World Forum on Internet of Things (WF-IoT)* 2015 Dec 14 (pp. 640–4). IEEE.

30. Yazdi M. Integration of IoT and edge computing in industrial systems. In *Advances in Computational Mathematics for Industrial System Reliability and Maintainability* 2024 Feb 25 (pp. 121–37). Springer Nature Switzerland.

31. Arena M, Azzone G, Conte A. A streamlined LCA framework to support early decision making in vehicle development. *Journal of Cleaner Production*, 2013 Feb 1;41:105–13.

32. Li H, Yazdi M. Advanced decision-making methods and applications in system safety and reliability problems. *Studies in Systems, Decision and Control* (vol. 211). Springer; 2022.

33. Yazdi M, editor. *Progressive Decision-making Tools and Applications in Project and Operation Management: Approaches, Case Studies, Multi-Criteria Decision-Making, Multi-Objective Decision-Making, Decision Under Uncertainty*. Springer; 2024.

34. Yazdi M, Khan F, Abbassi R, Rusli R. Improved DEMATEL methodology for effective safety management decision-making. *Safety Science*, 2020 July 1;127:104705.

35. Li H, Yazdi M. What are the critical well-drilling blowouts barriers? A progressive DEMATEL-game theory. In *Advanced Decision-Making Methods and Applications in System Safety and Reliability Problems: Approaches, Case Studies, Multi-Criteria Decision-Making, Multi-Objective Decision-Making, Fuzzy Risk-Based Models* 2022 July 10 (pp. 29–46). Springer International Publishing.

36. Heath RL, Abel DD. Proactive response to citizen risk concerns: Increasing citizens' knowledge of emergency response practices. *Journal of Public Relations Research*, 1996 July 1;8(3):151–71.

37. Tait J, Levidow L. Proactive and reactive approaches to risk regulation: The case of biotechnology. *Futures*, 1992 Apr 1;24(3):219–31.

38. Xu H, Kwan CM, Haynes L, Pryor JD. Real-time adaptive on-line traffic incident detection. *Fuzzy Sets and Systems*, 1998 Jan 16;93(2):173–83.

39. Li L, Lin Y, Du B, Yang F, Ran B. Real-time traffic incident detection based on a hybrid deep learning model. *Transportmetrica A: Transport Science*, 2022 Mar 4;18(1):78–98.

40. Rizvi SM, Ahmed A, Shen Y. Real-time incident detection and capacity estimation using loop detector data. *Journal of Advanced Transportation*, 2020;2020(1):8857502.

41. Poulymenopoulou M, Malamateniou F, Vassilacopoulos G. Emergency healthcare process automation using mobile computing and cloud services. *Journal of Medical Systems*, 2012 Oct;36:3233–41.

42. Clawson J, Olola CH, Heward A, Scott G, Patterson B. Accuracy of emergency medical dispatchers' subjective ability to identify when higher dispatch levels are warranted over a Medical Priority Dispatch System automated protocol's recommended coding based on paramedic outcome data. *Emergency Medicine Journal*, 2007 Aug 1;24(8):560–3.

43. Adesina KA, Yazdi M, Omidvar M. Emergency decision making fuzzy-expert aided disaster management system. In *Linguistic Methods Under Fuzzy Information in System Safety and Reliability Analysis* 2022 Mar 11 (pp. 139–50). Springer International Publishing.

44. Mendonca D, Beroggi GE, Wallace WA. Decision support for improvisation during emergency response operations. *International Journal of Emergency Management*, 2001 Jan 1;1(1):30–8.

45. Kondaveti R, Ganz A. Decision support system for resource allocation in disaster management. In *2009 Annual International Conference of the IEEE Engineering in Medicine and Biology Society* 2009 Sep 3 (pp. 3425–8). IEEE.

46. Cheng N, Yang L, Moradi R, Yazdi M. Empowering emergency operations management: A pride day. In *Progressive Decision-Making Tools and Applications in Project and Operation Management: Approaches, Case Studies, Multi-Criteria Decision-Making, Multi-Objective Decision-Making, Decision under Uncertainty* 2024 Mar 8 (pp. 109–20). Springer Nature Switzerland.

47. Dwivedi YK, Hughes L, Ismagilova E, Aarts G, Coombs C, Crick T, Duan Y, Dwivedi R, Edwards J, Eirug A, Galanos V. Artificial Intelligence (AI): Multidisciplinary perspectives on emerging challenges, opportunities, and agenda for research, practice and policy. *International Journal of Information Management*, 2021 Apr 1;57:101994.

48. Duan Y, Edwards JS, Dwivedi YK. Artificial intelligence for decision making in the era of Big Data – Evolution, challenges and research agenda. *International Journal of Information Management*, 2019 Oct 1;48:63–71.

49. Bayram V. Optimization models for large scale network evacuation planning and management: A literature review. *Surveys in Operations Research and Management Science*, 2016 Dec 1;21(2):63–84.

50. Lindell MK, Murray-Tuite P, Wolshon B, Baker EJ. *Large-Scale Evacuation: The Analysis, Modeling, and Management of Emergency Relocation from Hazardous Areas*. CRC Press; 2018 Dec 7.

5 Navigating the Future

Technological Advancements and Implementation Challenges in Workplace Health and Safety

5.1 THE HORIZON OF INNOVATION: EMERGING TECHNOLOGIES RESHAPING WORKPLACE HEALTH AND SAFETY

This section meticulously examines the forefront of technological innovation and its potential to dramatically reshape the field of Workplace Health and Safety (WHS). Two technologies stand out for their transformative capabilities: augmented reality (AR) and advanced robotics [1–2]. These technologies are additions to the WHS toolkit and pivotal in evolving the paradigm through which workplace safety is viewed and implemented [3–4].

5.1.1 AUGMENTED REALITY (AR)

AR technology is poised to revolutionize WHS by enhancing the way critical information is conveyed and interacted with within various working environments. This technology integrates digital data into the physical world in real time, allowing for an interactive experience that merges the real with the virtual. AR can provide workers with immediate access to vital safety data overlaid directly onto their field of view. For example, when equipped with AR headsets, workers can receive contextual safety warnings and instructions that adapt to their immediate surroundings, thereby significantly reducing the risk of accidents and enhancing compliance with stringent safety protocols.

Furthermore, AR's utility extends beyond immediate job site applications to include comprehensive training modules [5–6]. It offers a simulated yet highly realistic exposure to hazardous work conditions without the actual risks associated with these environments. Trainees can engage in these simulations, which can adjust dynamically to their responses and providing instant feedback. This tailored training

DOI: 10.1201/9781003515173-5

approach heightens understanding and retention of safety practices and ensures that workers are better prepared for a wide array of scenarios they might encounter.

5.1.2 ADVANCED ROBOTICS

The integration of robotics in WHS represents a monumental shift in how tasks perceived as dangerous or overly strenuous for humans are approached [7–8]. Robots are ideally suited to assume roles that involve high-risk environments—such as those characterized by extreme temperatures, the presence of toxic substances, or the requirement for work at great heights [9–10]. In these scenarios, robotics technology can prevent potential human injuries by taking on hazardous tasks with precision and efficiency.

Moreover, robots are invaluable in performing repetitive tasks that could lead to occupational injuries among human workers, such as musculoskeletal disorders resulting from repeated motions. As robotics technology continues to advance, particularly with the integration of artificial intelligence, these machines are performing isolated tasks and becoming capable of detecting potential hazards and initiating preventive measures in real time, thereby offering a proactive approach to workplace safety.

Despite the clear benefits, the adoption of AR and advanced robotics in WHS is fraught with challenges. These range from the high costs associated with implementing advanced technologies to resistance within the workforce, often due to fears of job displacement or the steep learning curve associated with new tools. To navigate these hurdles effectively, organizations must devise comprehensive strategies that include substantial investments in the necessary infrastructure, targeted training programs to equip staff with the skills to leverage these new technologies, and a concerted effort to foster an organizational culture that values and encourages innovation and ongoing improvement.

Table 5.1 categorizes and summarizes the impact of AR and advanced robotics on and WHS, paired with detailed case examples that illustrate these technologies in action.

TABLE 5.1
Technological Innovations in Workplace Health and Safety: Applications, Benefits, and Challenges

Technology	Application in WHS	Benefits	Challenges
Augmented Reality (AR)	Training and Real-time Information	Enhances learning, provides context-specific safety instructions	Cost, adaptation to existing workflows
Advanced Robotics	Hazardous Task Automation	Reduces human exposure to dangerous conditions, decreases injury rates	Initial investment, maintenance costs

5.1.3   CASE EXAMPLES

5.1.3.1   Case Example 1: Augmented Reality for Training in the Construction Industry

Context: A large construction firm implemented an AR-based training program to enhance safety for their workers. This involved the use of AR headsets that overlay safety protocols and hazard information directly onto the construction site [11–13].

Implementation: Trainees wore AR headsets that displayed real-time data as they navigated the construction site [14–15]. The system provided visual warnings and instructions, such as identifying load-bearing walls or indicating safe paths through the site.

Outcome: The use of AR led to a 40% reduction in training-related incidents and a 30% improvement in retention of safety procedures [16–17]. Feedback from employees suggested increased confidence in handling site-specific hazards, demonstrating the effectiveness of immersive, interactive training environments.

Challenges: The main challenges included the high cost of AR equipment and the need to update safety content regularly to reflect real time changes in the construction environment.

5.1.3.2   Case Example 2: Advanced Robotics in Chemical Manufacturing

Context: A chemical manufacturing company introduced robotics to handle the transport and mixing of hazardous materials, tasks previously performed by human workers.

Implementation: Robots designed for high precision and operation in corrosive environments were deployed to manage the transport, measurement, and mixing of reactive chemicals. These robots were equipped with sensors to detect leaks and malfunctions early.

Outcome: The automation of these hazardous tasks led to a significant decrease in human exposure to harmful substances and a drop in incident rates by over 50%. Additionally, the precision of robots reduced material waste during the mixing process, leading to cost savings.

Challenges: The company faced substantial upfront costs in purchasing and customizing robots for specific tasks. Ongoing expenses included maintenance and updates to robotic systems, as well as training for employees to operate and troubleshoot the automated equipment.

5.2   OVERCOMING OBSTACLES: ADDRESSING BARRIERS TO THE INTEGRATION OF AI AND DATA-DRIVEN TECHNOLOGIES IN WORKPLACE HEALTH AND SAFETY

This section meticulously examines the multifaceted and intricate challenges that obstruct the adoption of AI and data-driven technologies in the WHS areas [19–20]. It begins by dissecting the technological readiness required for integrating such advanced tools into existing WHS frameworks, highlighting the discrepancies between current practices and the capabilities afforded by cutting-edge technology [20–21]. This analysis is further deepened by exploring the complexities involved in harmonizing new technologies with established safety protocols and the significant adjustments needed within organizational structures and processes [22–24].

Moreover, the text addresses the substantial cultural resistance often encountered from personnel at various levels of an organization, who may exhibit skepticism or fear toward AI and machine learning systems. The discussion extends to include the ethical and legal quandaries that these technologies introduce, such as concerns regarding data privacy, the security of sensitive information, and the accountability for decisions made by automated systems, especially in scenarios where safety is at stake.

The narrative also contemplates the economic implications, considering the initial investment costs and the ongoing expenses associated with maintaining state-of-the-art technological solutions. It offers a thorough examination of strategic methodologies to surmount these barriers, advocating for a phased approach that includes pilot testing, comprehensive training programs tailored to enhance user confidence and proficiency, and continuous dialogue with stakeholders to ensure alignment with the organization's safety culture [25–27].

To round out the discussion, the section proposes the establishment of a robust governance framework that supports ethical practices and compliance with regulatory requirements, thereby fostering a secure environment for the integration of AI and data-driven technologies [28–29]. Through this exhaustive exploration, the section sheds light on the impediments to technological adoption in WHS and furnishes readers with practical, actionable strategies designed to facilitate a smoother and more effective integration of these technologies, ultimately enhancing safety outcomes in workplace environments [30–31].

Tables 5.2–5.5 are designed as a strategic resource for organizations planning to adopt AI and data-driven technologies in the domain of WHS. It is segmented into four distinct tabs, each focusing on a major category of barriers that typically hinder technology integration, alongside practical strategies to overcome these challenges:

1. **Technological Readiness**: This tab explores the technical prerequisites necessary for integrating AI and data-driven technologies [32–33]. It addresses factors such as infrastructure adequacy, system integration challenges, and the reliability of the technology. Each barrier is paired with an example and a suggested mitigation strategy to ensure seamless technological integration.
2. **Cultural Acceptance**: This section talks about the human aspect of technology adoption, highlighting the importance of employee trust, effective change management, and leadership support [34–35]. It suggests strategies such as educational workshops, gradual implementation processes, and leadership-driven communication to enhance acceptance and smooth the transition.
3. **Legal and Ethical Considerations**: This tab outlines the legal and ethical issues associated with deploying AI and data technologies in WHS [36–37]. It discusses the protection of data privacy, accountability for AI-driven decisions, and compliance with existing regulations [38–39]. Strategies

proposed include implementing robust cybersecurity measures, establishing clear guidelines for AI usage, and conducting regular legal audits.

4. **Economic Factors**: The final tab addresses the financial implications of technology adoption, focusing on the initial costs, the calculation of return on investment (ROI), and ongoing maintenance expenses [40–43]. It provides guidance on navigating financial challenges through incentives, clear ROI tracking, and sustainable financial planning.

Each tab serves as a comprehensive guide, helping organizations navigate the complexities of technology adoption in WHS by providing actionable solutions to common challenges. This format facilitates a detailed understanding of each barrier and equips stakeholders with the tools needed to foster a supportive environment for technological advancement in safety practices.

TABLE 5.2
Technological Readiness

Factor	Description	Example Challenge	Mitigation Strategy
Infrastructure	Adequacy of existing IT infrastructure to support new technologies.	Inadequate server capacity	Upgrade servers and invest in scalable cloud solutions.
Integration	Compatibility of new technologies with existing systems.	Mismatched data formats	Employ middleware or use APIs for seamless integration.
Reliability	Consistency and dependability of technology performance.	Frequent system downtimes	Implement robust testing phases to ensure technology reliability.

TABLE 5.3
Cultural Acceptance

Factor	Description	Example Challenge	Mitigation Strategy
Employee Trust	Trust in AI and data-driven decisions among staff.	Resistance to AI decisions	Conduct workshops to demonstrate the accuracy and safety of AI.
Change Management	Managing how changes due to technology adoption are handled.	Disruption in workflow	Provide comprehensive training and gradual implementation phases.
Leadership Support	Active support from top management for technology adoption.	Lack of executive buy-in	Develop clear communication from leadership emphasizing benefits.

TABLE 5.4
Legal and Ethical Considerations

Factor	Description	Example Challenge	Mitigation Strategy
Data Privacy	Protection of personal and sensitive data.	Risk of data breaches	Implement state-of-the-art cybersecurity measures.
Accountability	Determining responsibility for decisions made by AI.	Ambiguity in AI decision-making	Establish clear guidelines for AI decision-making processes.
Compliance	Adherence to legal standards and regulations.	Non-compliance risks	Regular audits and updates to technology based on legal advice.

TABLE 5.5
Economic Factors

Factor	Description	Example Challenge	Mitigation Strategy
Initial Costs	Initial investment required for technology acquisition and implementation.	High upfront costs	Seek financial incentives, subsidies, or phased investment plans.
ROI Calculation	Demonstrating the return on investment for technology spending.	Unclear ROI projections	Develop metrics and KPIs to track performance and cost savings.
Maintenance Costs	Ongoing expenses for maintaining and updating technology.	Costly software updates	Plan for long-term financial sustainability in technology planning.

5.3 BLUEPRINT FOR INTEGRATION: COMPREHENSIVE STRATEGIES FOR SURMOUNTING BARRIERS AND EFFECTIVELY IMPLEMENTING AI AND DATA-DRIVEN TECHNOLOGIES IN WORKPLACE HEALTH AND SAFETY

This section serves as an essential blueprint for organizations aspiring to integrate AI and data-driven technologies into their WHS systems. It methodically unpacks a spectrum of strategic interventions designed to navigate and overcome the multifaceted challenges associated with such technological adoptions. The discourse begins by identifying key barriers—ranging from infrastructural inadequacies and resistance within corporate culture to stringent legal constraints and significant economic considerations—that often impede the seamless integration of innovative technologies into existing safety frameworks.

To address these hurdles, the section proposes a layered approach, combining tactical, organizational, and regulatory strategies. On the tactical level, it advocates for the initiation of pilot projects to demonstrate tangible benefits, thereby fostering a culture of trust and acceptance among employees. These pilots can also serve as a test bed for refining technological functionalities and integration processes before a full-scale rollout.

At the organizational level, the narrative emphasizes the critical role of leadership in championing the adoption of these technologies. It suggests that executives and senior managers undertake visible roles in advocating for and reinforcing the value of AI and data-driven solutions [43–44], thus catalyzing a shift in organizational norms and attitudes. Furthermore, it recommends the establishment of cross-functional teams that include IT specialists, WHS personnel, and frontline workers to ensure that the implementation process is inclusive and considers diverse perspectives and needs. Regulatory strategies are also highlighted, underscoring the importance of aligning technological implementations with existing legal frameworks and ethical standards. This includes rigorous compliance with data protection laws, the development of transparent protocols for AI decision-making, and continuous monitoring to adapt to evolving regulatory landscapes [45–47]. Economically, the section provides a nuanced analysis of cost management, suggesting strategies such as leveraging government grants, seeking partnerships with technology providers for cost-sharing opportunities, and conducting comprehensive cost-benefit analyses to ensure sustainable investments.

Finally, the discourse wraps up by proposing continuous education and training programs as foundational to successful technology adoption [48–49]. These initiatives should be tailored to enhance skill levels across all employee strata and updated regularly to keep pace with technological advancements, thereby ensuring that the workforce is prepared to use new technologies and proficient in optimizing their benefits. Through this expansive and detailed examination, the section equips readers with a robust set of strategies, fostering a well-rounded understanding of the necessary steps to effectively implement AI and data-driven technologies in WHS, thereby transforming challenges into opportunities for innovation and enhanced safety.

Table 5.6 presents framework for organizations seeking to effectively integrate AI and data-driven technologies into their WHS systems. It is meticulously designed to outline key strategies across several crucial domains—Tactical, Organizational, Regulatory, Economic, and Education & Training—each essential for navigating the complex landscape of technology adoption.

1. **Tactical Domain**: This section focuses on practical, immediate steps like pilot projects and iterative implementations, providing organizations with a roadmap to initiate and refine the integration process [50–51]. These strategies ensure that the introduction of new technologies is manageable and their impacts measurable, allowing for adjustments before full-scale deployment.
2. **Organizational Domain**: It emphasizes the importance of leadership endorsement and the formation of cross-functional teams [52–53]. With securing support from top management and fostering collaboration across departments,

TABLE 5.6
Strategies for Effective Integration of AI and Data-Driven Technologies in WHS

Domain	Strategy	Description	Examples
Tactical	Pilot Projects	Implement small-scale projects to test and demonstrate technology benefits.	Starting with a pilot project in a single department to assess the impact of AI on incident reporting efficiency.
	Iterative Implementation	Gradually introduce technology in phases to ensure smooth integration.	Phased rollout of data analytics tools for risk assessment, starting with high-risk areas.
Organizational	Leadership Endorsement	Encourage senior management to advocate for and support technology adoption.	Senior managers publicly sharing success stories from the pilot projects to inspire wider acceptance.
	Cross-functional Teams	Form teams from different departments to facilitate inclusive implementation.	Creating a task force with IT, WHS, and operations staff to oversee the deployment of a new safety monitoring system.
Regulatory	Compliance with Legal Standards	Ensure all technologies adhere to current laws and ethical guidelines.	Regular reviews and updates of AI systems to ensure compliance with the latest data protection regulations.
	Transparent AI Protocols	Develop clear guidelines for the operations of AI systems.	Establishing a protocol for AI decision-making that includes checks for fairness and accountability.
Economic	Leverage Funding Opportunities	Utilize financial aids like grants or partnerships to mitigate costs.	Applying for a government grant to fund the initial setup of sensor-based monitoring systems in hazardous work areas.

TABLE 5.6 (Continued)
Strategies for Effective Integration of AI and Data-Driven
Technologies in WHS

Domain	Strategy	Description	Examples
	Cost-Benefit Analysis	Regularly evaluate the financial impact versus the benefits of new technologies.	Conducting semi-annual reviews to assess the ROI from automated equipment monitoring systems.
Education & Training	Continuous Learning Programs	Provide ongoing education and training for all levels of staff.	Implementing a quarterly workshop on the latest AI tools in safety management for all employees.
	Skill Enhancement Initiatives	Focus on improving specific skills related to new technologies.	Special training sessions for data analysts in the WHS department on interpreting AI-generated safety reports.

organizations can enhance buy-in and ensure diverse input in technology deployment, leading to more effective and inclusive implementations.

3. **Regulatory Domain**: This part of the table addresses compliance with legal standards and the establishment of transparent protocols for AI operations. It ensures that technological adoptions are effective, ethical, and lawful, safeguarding the organization against potential legal and ethical pitfalls.

4. **Economic Domain**: Strategies here focus on managing the financial aspects of technology adoption. Leveraging funding opportunities and conducting cost-benefit analyses help organizations maintain fiscal responsibility while investing in advanced technologies, ensuring that the adoption is financially sustainable.

5. **Education and Training Domain**: Recognizing that the successful implementation of new technologies requires a knowledgeable and skilled workforce [54–55], this domain advocates for continuous learning programs and skill enhancement initiatives. These strategies aim to prepare and empower employees at all levels to effectively use and benefit from the newly integrated technologies.

Each strategy in the table is complemented by practical examples that illustrate how these approaches can be implemented in real-world settings. This detailed structure aids in understanding the multifaceted process of technology integration and provides actionable steps that organizations can follow to ensure a smooth and effective transition to a more technologically advanced WHS environment.

5.4 FUTURE HORIZONS: ENVISIONING THE EVOLUTION OF WORKPLACE HEALTH AND SAFETY IN THE AGE OF TECHNOLOGICAL INNOVATION

This section embarks on a visionary exploration of the future landscape of WHS as it intertwines with the relentless march of technological advancements. It posits a forward-looking analysis, delving into how emerging technologies such as AI, the internet of things (IoT), and Big Data analytics will revolutionize WHS practices. The narrative begins by projecting the potential transformative impacts of these technologies on risk assessment, hazard detection, and incident response, suggesting a shift toward more predictive and preventative safety measures [56–58].

As we peer into the future, the discussion expands to consider the integration of sophisticated wearable devices and sensors that could provide real-time health monitoring and environmental scanning, thereby enhancing the hazard identification precision and the timeliness of interventions. This segment hypothesizes about the next-generation safety protocols that could become standard practice, such as dynamic risk assessment tools that adapt to changing conditions and machine learning models that evolve from past incidents [59–61].

Furthermore, the text speculates on the role of augmented and virtual reality in training and simulation, offering scenarios where employees can engage in immersive, realistic safety training modules without the risks associated with physical training environments. This could particularly revolutionize training in high-risk industries, providing a safe yet effective method for preparing workers for hazardous situations.

Ethical considerations and challenges are also scrutinized, particularly the implications of increased surveillance and data collection, which, while beneficial for safety, might raise concerns about privacy and autonomy. The section discusses the balance needed between leveraging technology for safety improvements and respecting individual rights, suggesting that future regulations and ethical guidelines will need to evolve alongside technological innovations.

The prospect of global standardization of safety practices facilitated by universally accessible technologies is considered. This could lead to a more uniform approach to safety across borders, benefiting industries that operate internationally by simplifying compliance and enhancing overall safety standards.

Through this comprehensive and speculative analysis, the section aims to provide a well-rounded view of the potential futures of WHS, equipped with both the promises and challenges brought about by technological advancements. It serves as a thought-provoking piece for stakeholders to consider as they plan, highlighting the need for adaptive strategies that embrace innovation while maintaining the core goal of protecting worker health and safety.

Exploring potential challenges in the integration of advanced technologies into WHS systems, along with practical examples, can provide valuable insights into the complexities involved. Here are some significant challenges along with scenarios that illustrate these issues:

1. **Data Privacy and Security**:
   - **Challenge**: As WHS systems increasingly rely on personal and sensitive data collected through various technologies like wearables and sensors, ensuring the security and privacy of this data becomes paramount.
   - **Example**: A manufacturing company introduces smart helmets that monitor the physiological states of workers to prevent heat stress and fatigue. However, the company must address concerns regarding the unauthorized access and potential misuse of workers' health data, requiring robust encryption and strict access controls.
2. **Technology Adoption and User Resistance**:
   - **Challenge**: Resistance from employees, often stemming from fears of job displacement or mistrust of new technology, can hinder the adoption of innovative solutions.
   - **Example**: A logistics firm implements an AI-driven system designed to optimize warehouse operations and enhance safety protocols. Despite its potential, the initiative faces pushback from workers who are concerned about surveillance and the reduction of workforce due to automation. To mitigate this, the firm initiates a series of engagement workshops to educate employees on the benefits of the system and how it actually aims to assist rather than replace them.
3. **Cost of Implementation**:
   - **Challenge**: The initial investment for cutting-edge technology can be prohibitive for many organizations, particularly small to medium-sized enterprises (SMEs).
   - **Example**: A small construction company struggles with the high costs associated with deploying drones for real-time surveillance of construction sites to monitor safety compliance. The company could explore financial support options like technology grants or partnerships with tech firms offering pilot programs at a reduced cost.
4. **Regulatory Compliance**:
   - **Challenge**: Keeping up with changing regulations that govern the use of technologies in the workplace can be challenging, especially when these technologies evolve faster than the legislative framework.
   - **Example**: A healthcare provider uses AI algorithms to predict workplace injuries but finds it challenging to comply with both healthcare privacy laws and workplace safety regulations, which differ significantly. Regular consultations with legal experts and ongoing training for compliance officers are necessary to navigate these regulatory landscapes.
5. **Integration with Existing Systems**:
   - **Challenge**: Integrating new technologies with existing WHS management systems without disrupting current operations can be difficult.

- **Example**: An oil and gas company introduces a real-time data analytics platform to predict equipment failure and potential safety hazards. The integration phase encounters numerous technical issues as the new platform struggles to synchronize data with older, legacy safety management systems [62–63]. The solution involves incremental integration and possibly employing specialized IT consultants to ensure smooth data flow and system functionality.

6. **Maintenance and Technological Obsolescence**:
   - **Challenge**: Ensuring that technological solutions remain up-to-date and maintaining them can be costly and require technical expertise not always available in-house.
   - **Example**: An industrial plant installs an advanced sensor network to monitor toxic gas levels. However, the rapid pace of technological advancement renders the sensors obsolete within a few years, necessitating another significant investment. Adopting technology as a service (TaaS) could be a solution, where updates and maintenance are managed by the provider for a regular fee.

Addressing these challenges requires a proactive approach, thoughtful planning, and often creative financial and operational strategies to ensure that the integration of technology into WHS practices is both successful and sustainable.

5.5 CONCLUSION

As we conclude this exploration into the intersection of emerging technologies and WHS, it is evident that the future holds profound transformative potential for the field. Technologies such as AR, advanced robotics, and AI are auxiliary tools, and fundamental catalysts poised to redefine safety practices across industries. These innovations offer significant benefits, including enhanced training capabilities, improved risk assessments, and the automation of hazardous tasks, which collectively contribute to safer and more efficient workplaces.

However, the journey toward full integration of these technologies is laden with substantial challenges. High implementation costs, cultural resistance, and the complexities of marrying new systems with existing infrastructures are significant hurdles that organizations must navigate. To me, addressing these challenges necessitates a holistic approach that extends beyond technological adoption to include strategic planning, cultural adaptation, and continuous education.

Organizations must foster a culture that embraces change and innovation while ensuring that technological advancements align with the core goal of WHS— protecting worker health and safety. This requires substantial investments in technology and in the human elements of training and support systems that encourage acceptance and optimize the use of new tools.

In moving forward, it is crucial for stakeholders to engage in continuous dialogue, pilot testing, and phased implementation strategies. These steps will facilitate smoother integration and help tailor technological solutions to meet specific organizational needs. Additionally, maintaining an agile approach to technology adoption—ready to adapt to new developments and changing regulations—is essential for keeping pace with the rapid evolution of both technology and workplace requirements.

By actively addressing these challenges and leveraging the opportunities provided by technological advancements, we can anticipate a future where WHS practices are reactive and predominantly predictive and preventive. This shift will undoubtedly lead to a significant reduction in workplace incidents and injuries, ultimately fostering a safer work environment for all.

As we navigate the future of WHS, our success will largely depend on our ability to integrate advanced technologies with human-centric strategies. This approach will ensure that the benefits of innovation are fully realized, transforming potential challenges into opportunities for enhancement and growth in workplace safety.

REFERENCES

1. Javornik A, Marder B, Pizzetti M, Warlop L. Augmented self-The effects of virtual face augmentation on consumers' self-concept. *Journal of Business Research*, 2021 June 1;130:170–87.
2. Benyon D. Presence in blended spaces. *Interacting with Computers*, 2012 July;24(4):219–26.
3. Paulson SK. *"Occupational Health and Safety" Corporate Liability and the Regulation of Officers: New Zealand Reform*. (Doctoral dissertation). Open Access Te Herenga Waka-Victoria University of Wellington.
4. Yazdi M. Integration of computational mathematics in industrial decision-making. In *Advances in Computational Mathematics for Industrial System Reliability and Maintainability* 2024 Feb 25 (pp. 105–20). Springer Nature Switzerland.
5. Carmigniani J, Furht B, Anisetti M, Ceravolo P, Damiani E, Ivkovic M. Augmented reality technologies, systems and applications. *Multimedia Tools and Applications*, 2011 Jan;51:341–77.
6. Azuma R, Baillot Y, Behringer R, Feiner S, Julier S, MacIntyre B. Recent advances in augmented reality. *IEEE Computer Graphics and Applications*, 2001 Nov;21(6):34–47.
7. Khan MY, Zaina F, Muhammad S, Tapete D. Integrating Copernicus Satellite products and ground-truthing for documenting and monitoring the impact of the 2022 extreme floods in Pakistan on cultural heritage. *Remote Sensing*, 2023 May 10;15(10):2518.
8. Goussous JS. Artificial intelligence-based restoration: The case of Petra. *Civil Engineering and Architecture*, 2020 Dec;8(6):1350–8.
9. Cherrie JW, Semple S, Christopher Y, Saleem A, Hughson GW, Philips A. How important is inadvertent ingestion of hazardous substances at work? *The Annals of Occupational Hygiene*, 2006 Oct 1;50(7):693–704.
10. Carson PA. *Hazardous Chemicals Handbook*. Elsevier; 2002 Mar 12.
11. Li X, Yi W, Chi HL, Wang X, Chan AP. A critical review of virtual and augmented reality (VR/AR) applications in construction safety. *Automation in Construction*, 2018 Feb 1;86:150–62.
12. Pierdicca R, Prist M, Monteriù A, Frontoni E, Ciarapica F, Bevilacqua M, Mazzuto G. Augmented reality smart glasses in the workplace: Safety and security in the fourth industrial revolution era. In *Augmented Reality, Virtual Reality, and Computer Graphics: 7th International Conference, AVR 2020*, Lecce, Italy, September 7–10, 2020, Proceedings, Part II 7 2020 (pp. 231–47). Springer International Publishing.

13. Yazdi M. Augmented reality (AR) and virtual reality (VR) in maintenance training. In *Advances in Computational Mathematics for Industrial System Reliability and Maintainability* 2024 Feb 25 (pp. 169–83). Springer Nature Switzerland.
14. Behzadan AH, Dong S, Kamat VR. Augmented reality visualization: A review of civil infrastructure system applications. *Advanced Engineering Informatics*, 2015 Apr 1;29(2):252–67.
15. Casini M. Extended reality for smart building operation and maintenance: A review. *Energies*, 2022 May 20;15(10):3785.
16. Chiang FK, Shang X, Qiao L. Augmented reality in vocational training: A systematic review of research and applications. *Computers in Human Behavior*, 2022 Apr 1;129:107125.
17. Cohen A, Colligan MJ. *Assessing Occupational Safety and Health Training*. National Institute for Occupational Safety and Health; 1998.
18. Cebulla A, Szpak Z, Knight G. Preparing to work with artificial intelligence: Assessing WHS when using AI in the workplace. *International Journal of Workplace Health Management*, 2023 July 28;16(4):294–312.
19. Yazdi M, Zarei E, Adumene S, Beheshti A. Navigating the power of artificial intelligence in risk management: A comparative analysis. *Safety*, 2024 Apr 26;10(2):42.
20. Ameen N, Hosany S, Tarhini A. Consumer interaction with cutting-edge technologies: Implications for future research. *Computers in Human Behavior*, 2021 July 1;120:106761.
21. Arduini F, Cinti S, Scognamiglio V, Moscone D, Palleschi G. How cutting-edge technologies impact the design of electrochemical (bio) sensors for environmental analysis. A review. *Analytica Chimica Acta*, 2017 Mar 22;959:15–42.
22. Li Y, Guldenmund FW. Safety management systems: A broad overview of the literature. *Safety Science*, 2018 Mar 1;103:94–123.
23. Slikker Jr W, de Souza Lima TA, Archella D, de Silva Junior JB, Barton-Maclaren T, Bo L, Buvinich D, Chaudhry Q, Chuan P, Deluyker H, Domselaar G. Emerging technologies for food and drug safety. *Regulatory Toxicology and Pharmacology*, 2018 Oct 1;98:115–28.
24. Yazdi M. Enhancing system safety and reliability through integrated FMEA and game theory: A multi-factor approach. *Safety*, 2023 Dec 22;10(1):4.
25. Mohamed S. Scorecard approach to benchmarking organizational safety culture in construction. *Journal of Construction Engineering and Management*, 2003 Feb;129(1):80–8.
26. Olian JD, Rynes SL. Making total quality work: Aligning organizational processes, performance measures, and stakeholders. *Human Resource Management*, 1991 Sep;30(3):303–33.
27. Yazdi M, Khan F, Abbassi R, Rusli R. Improved DEMATEL methodology for effective safety management decision-making. *Safety Science*, 2020 July 1;127:104705.
28. Johnson M, Jain R, Brennan-Tonetta P, Swartz E, Silver D, Paolini J, Mamonov S, Hill C. Impact of big data and artificial intelligence on industry: Developing a workforce roadmap for a data driven economy. *Global Journal of Flexible Systems Management*, 2021 Sep;22(3):197–217.
29. Ahmad K, Iqbal W, El-Hassan A, Qadir J, Benhaddou D, Ayyash M, Al-Fuqaha A. Data-driven artificial intelligence in education: A comprehensive review. *IEEE Transactions on Learning Technologies*, 2024;17:12–31.
30. Fraboni F, Brendel H, Pietrantoni L. Evaluating organizational guidelines for enhancing psychological well-being, safety, and performance in technology integration. *Sustainability*, 2023 May 16;15(10):8113.

31. Robla-Gómez S, Becerra VM, Llata JR, Gonzalez-Sarabia E, Torre-Ferrero C, Perez-Oria J. Working together: A review on safe human-robot collaboration in industrial environments. *IEEE Access*, 2017 Nov 14;5:26754–73.

32. Li XH, Cao CC, Shi Y, Bai W, Gao H, Qiu L, Wang C, Gao Y, Zhang S, Xue X, Chen L. A survey of data-driven and knowledge-aware explainable AI. *IEEE Transactions on Knowledge and Data Engineering*, 2020 Mar 30;34(1):29–49.

33. Bechtsis D, Tsolakis N, Iakovou E, Vlachos D. Data-driven secure, resilient and sustainable supply chains: Gaps, opportunities, and a new generalised data sharing and data monetisation framework. *International Journal of Production Research*, 2022 July 18;60(14):4397–417.

34. Davis FD, Bagozzi RP, Warshaw PR. Technology acceptance model. *Journal of Management and Science*, 1989;35(8):982–1003.

35. Venkatesh V, Davis F, Morris MG. Dead or alive? The development, trajectory and future of technology adoption research. *Journal of the AIS* 2007 Apr 27;8(1):267–86.

36. Gunn J, Taylor P. *Forensic Psychiatry: Clinical, Legal and Ethical Issues*. CRC Press; 2014 Jan 6.

37. Nagappan A, Kalokairinou L, Wexler A. Ethical and legal considerations of alternative neurotherapies. *AJOB Neuroscience*, 2021 Oct 2;12(4):257–69.

38. Raparthi M. AI-driven decision support systems for precision medicine: Examining the development and implementation of AI-driven decision support systems in precision medicine. *Journal of Artificial Intelligence Research*, 2021 Apr 12;1(1):11–20.

39. Araujo T, De Vreese C, Helberger N, Kruikemeier S, van Weert J, Bol N, Oberski D, Pechenizkiy M, Schaap G, Taylor L. Automated decision-making fairness in an AI-driven world: Public perceptions, hopes and concerns. *Digital Communication Methods Lab*, 2018 Sep 25.

40. Connolly MP, Hoorens S, Chambers GM. The costs and consequences of assisted reproductive technology: An economic perspective. *Human Reproduction Update*, 2010 Nov 1;16(6):603–13.

41. North D, Baldock R, Ullah F. Funding the growth of UK technology-based small firms since the financial crash: Are there breakages in the finance escalator? *Venture Capital*, 2013 July 1;15(3):237–60.

42. Li H, Peng W, Adumene S, Yazdi M. Operations management of critical energy infrastructure: A sustainable approach. In *Intelligent Reliability and Maintainability of Energy Infrastructure Assets* 2023 May 4 (pp. 39–52). Springer Nature Switzerland.

43. Henke N, Jacques Bughin L. *The Age of Analytics: Competing in a Data-Driven World*.

44. Jain P, Tripathi V, Malladi R, Khang A. Data-driven artificial intelligence (AI) models in the workforce development planning. In *Designing Workforce Management Systems for Industry 4.0* 2023 (pp. 159–76). CRC Press.

45. Ranger N, Millner A, Dietz S, Fankhauser S, Lopez A, Ruta G. Adaptation in the UK: A decision-making process. *Environment Agency*, 2010 Sep;9:1–62.

46. Lawrence JH. *The Adequacy of Institutional Frameworks and Practice for Climate Change Adaptation Decision Making* (Doctoral dissertation). Open Access Te Herenga Waka-Victoria University of Wellington.

47. Li H, Yazdi M. Advanced decision-making methods and applications in system safety and reliability problems. *Studies in Systems, Decision and Control*, 2022:211.

48. Straub ET. Understanding technology adoption: Theory and future directions for informal learning. *Review of Educational Research*, 2009 June;79(2):625–49.

49. National Research Council. *Division of Behavioral, Social Sciences, Board on Testing, Assessment, Board on Science Education, Committee on Highly Successful Schools or Programs for K-12 STEM Education. Successful K-12 STEM Education: Identifying*

Effective Approaches in Science, Technology, Engineering, and Mathematics. National Academies Press; 2011 Jul 22.

50. Ballard G, Kim YW, Jang JW, Liu M. Road map for lean implementation at the project level. *The Construction Industry Institute*, 2007 Oct;11:234.

51. Brinkerhoff DW, Ingle MD. Integrating blueprint and process: A structured flexibility approach to development management. *Public Administration and Development*, 1989 Nov;9(5):487–503.

52. Major DA, Litano ML. The importance of organizational leadership in managing work and family. *The Oxford Handbook of Work and Family* 2016 May 17 (pp. 242–54).

53. Van Kleef GA, Heerdink MW, Cheshin A, Stamkou E, Wanders F, Koning LF, Fang X, Georgeac OA. No guts, no glory? How risk-taking shapes dominance, prestige, and leadership endorsement. *Journal of Applied Psychology*, 2021 Nov;106(11):1673.

54. Hutchins HM, Burke LA. Identifying trainers' knowledge of training transfer research findings – Closing the gap between research and practice. *International Journal of Training and Development*, 2007 Dec;11(4):236–64.

55. Hilton ML, Pellegrino JW, editors. *Education for Life and Work: Developing Transferable Knowledge and Skills in the 21st Century.* National Academies Press; 2012 Dec 18.

56. Elgendy N, Elragal A. Big data analytics: A literature review paper. In *Advances in Data Mining. Applications and Theoretical Aspects: 14th Industrial Conference, ICDM 2014, St. Petersburg, Russia, July 16–20, 2014.* Proceedings 14 2014 (pp. 214–27). Springer International Publishing.

57. Yazdi M. The application of bow-tie method in hydrogen sulfide risk management using layer of protection analysis (LOPA). *Journal of Failure Analysis and Prevention*, 2017 Apr;17:291–303.

58. Li H, Yazdi M. How to deal with toxic people using a fuzzy cognitive map: Improving the health and wellbeing of the human system. In *Advanced Decision-Making Methods and Applications in System Safety and Reliability Problems: Approaches, Case Studies, Multi-Criteria Decision-Making, Multi-Objective Decision-Making, Fuzzy Risk-Based Models* 2022 July 10 (pp. 87–107). Springer International Publishing.

59. Yazdi M, Golilarz NA, Nedjati A, Adesina KA. An improved lasso regression model for evaluating the efficiency of intervention actions in a system reliability analysis. *Neural Computing and Applications*, 2021 July;33(13):7913–28.

60. Yazdi M. A perceptual computing-based method to prioritize intervention actions in the probabilistic risk assessment techniques. *Quality and Reliability Engineering International*, 2020 Feb;36(1):187–213.

61. Adumene S, Okwu M, Yazdi M, Afenyo M, Islam R, Orji CU, Obeng F, Goerlandt F. Dynamic logistics disruption risk model for offshore supply vessel operations in Arctic waters. *Maritime Transport Research*, 2021 Jan 1;2:100039.

62. Easton C. Safety integrity verification of legacy systems. *Measurement and Control*, 2009 July;42(6):185–9.

63. Leveson NG. Rasmussen's legacy: A paradigm change in engineering for safety. *Applied Ergonomics*, 2017 Mar 1;59:581–91.

Index

A

Actionable insights-deriving from strategic analytics, 55, 57, 58, 62

Advanced robotics-applications in hazardous environments, 95–97

Advanced statistical methods-application in predictive analytics, 55

AI and data-driven technologies-integration challenges in WHS, 97, 98, 100, 102

AI deployment-in risk management, 88–90

AI technology-impact on organizational safety culture, 84

Analytical techniques-integration for risk evaluations, 61

Artificial intelligence (AI): benefits in workplace safety 75, 80, 96; predictive maintenance models, 9, 50

Augmented reality (AR)-enhancement of safety data presentation, 95, 97

Automation-impact on safety and efficiency, 24, 83

Automation strategies-in risk assessment, 79

B

Baseline safety conditions-establishing through historical data, 34

Big data analytics: in health care, 1; predictive safety management, 104

Biometric sensors-monitoring worker health, 2

C

Case examples-practical applications of AR and robotics in industries, 96

Case studies-showcasing analytics in various industries, 47, 48, 55, 72

Chemical manufacturing-use of robotics for handling hazardous materials, 97

Chemical spills-incident statistics and control measures, 35, 86

Cloud computing-data management and safety analysis, 18

Compliance-maintaining through descriptive analytics, 55

Compliance-shift from compliance-based to proactive data-driven approaches, 59

Compliance tracking-regular audits and legal adherence, 36, 37

Construction industry-AR for training and safety enhancements, 59, 64, 97

Cost and workforce challenges-barriers to implementing AR and robotics, 96, 97

Cultural acceptance-enhancing trust and management of technological changes, 98, 99

Cultural resistance-overcoming employee skepticism towards new technologies, 96, 106

D

Data analytics training-for safety officers, 37

Database utilization-for trend analysis and predictive analytics, 36

Data bias-addressing fairness in AI applications, 88, 90

Data-driven approaches-transformation of WHS, 8; proactive safety management, 41

Data integration-combining sources for comprehensive analysis, 26, 39, 42

Data privacy-management in workplace health and safety, 42, 47, 49, 98

Data protection-implementation of secure protocols, 38, 41, 47, 49

Data quality management-ensuring accuracy and completeness, 47–49

Data quality-improvements through technology, 24, 38, 47

Data sources-diverse contributions to safety assessments, 34

Data summarization-using descriptive analytics in safety management, 56, 58

Decision-making-enhancement through data-driven insights, 65, 83

Decision-making support-AI in crisis situations, 100

Descriptive analytics-best practices and operational impact, 55, 56

Digital twins-in aviation safety, 26

Drones-for structural assessments in construction, 10, 18, 19; aerial surveillance in agriculture, 105

E

Economic factors-addressing financial challenges in tech adoption, 100

Economic implications-cost management and ROI of technological solutions, 98

Education and training-continuous learning for effective tech use in WHS, 101, 103

Efficiency-operational improvements through data-driven technologies, 19